Rise *of the* RedMed

How the Mediterranean-Red Sea Nexus is Resuming its Strategic Centrality

Rise *of the* RedMed

How the Mediterranean-Red Sea Nexus is Resuming its Strategic Centrality

By

Gregory R. Copley
Lt.-Gen. Aliyu Gusau Mohammed (rtd.)
Prince Ermias Sahle-Selassie Haile-Selassie
Ambassador Princeton Lyman
Andrew Pickford
Pamela von Gruber

ISSA Gusau Institute Gi WIA

Rise *of the* RedMed

How the Mediterranean-Red Sea Nexus is Resuming its Strategic Centrality

ISSA

Published by

The International Strategic Studies Association
PO Box 320608,
Alexandria, Virginia 22320,
United States of America.
Telephone +1 (703) 548-1070.
Facsimile +1 (703) 684-7476.
Email: Marketing@StrategicStudies.org
www.StrategicStudies.org

for
The International Strategic Studies Association
The Gusau Institute
The Water Initiative for Africa

Manufactured in the United States of America.

First Edition

ISBN-13: 978-1-892998-24-8
ISBN-10: 1-892998-24-6

US$22.50 plus shipping and handling

THIS BOOK IS DEDICATED TO THE MEMORY OF

All of the Founding Fathers of the Organization of African Unity
(now the African Union),
and particularly to those OAU leaders who worked together to make the RedMed strategic zone an area of global importance:

His Imperial Majesty Emperor Haile Selassie I
of Ethiopia
and
President Gamal Abdul Nasser,
of Egypt

And to all of the People of the RedMed States.

If the King of Portugal had made himself master of Aden with a good fortress, such as those at Hormuz and Malacca, and so held sway over these three straits … he might well have been called lord of all the world … for with these three keys in his hands he might shut the doors against all comers.

— Admiral Alfonso d'Albuquerque, in the final of his *Commentaries*, 1576.

There is a tide in the affairs of men
Which, taken at the flood, leads on to fortune;
Omitted, all the voyage of their life
Is bound in shallows and in miseries.
On such a full sea are we now afloat,
And we must take the current when it serves,
Or lose our ventures.

— William Shakespeare: *Julius Cæsar*, Act IV, Scene ii.

Contents

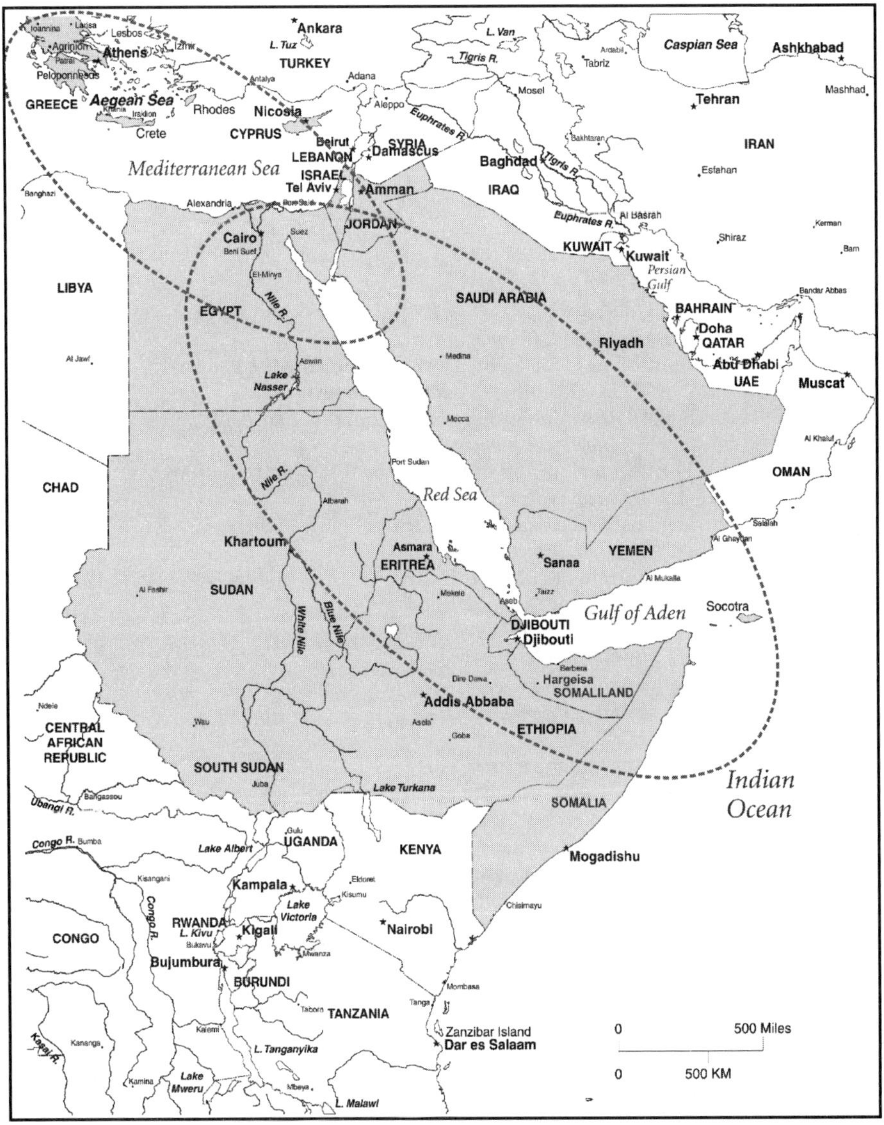

The RedMed combined theater countries shown shaded, with the upper dotted oval showing the north-western states and the lower dotted oval showing the south-eastern states.

Preface

An Area of Unusual Importance

History has always given primacy to that great conjunction of societies which grew from the Eastern Mediterranean and Nile Basin, linked to hinterlands in Arabia, the Persian Gulf, the Red Sea, and the Horn of Africa. These societies and cultures became the font of civilizations, even before they became significant as the nexus of east-west and north-south trade in the modern era.

Today, this complex of overlapping strategic theaters — which we are calling, in this book, "the RedMed" (the Red Sea and the Mediterranean) — is the sum of all of its past as a cradle of civilizations and as a nexus of trade ... plus something entirely new. That "something new" is the potential of the region to become the hub of a modern framework of new markets, including the arc of Mediterranean littoral states of Egypt, Israel, Greece, and Cyprus, down to the Red Sea states of Jordan, Saudi Arabia, Ethiopia, Djibouti, and Somaliland.

If. If certain conflicts can be overcome.

And When. When the current processes of national and societal transformations become crystalized over the coming decade or two.

This evolution, or coalescing, of the societies into revived civilizational structures and a potential new and vibrant common market could be the result of processes and prospects now deeply underway. Some of the societies in the RedMed have already begun to flourish according to a regional agenda which began to emerge with the loosening of the grip of superpower engagement in the region.

That loosening of superpower engagement began in the second decade of the 21st Century and saw, for the first time in some 500 years, a relative absence of external power dominance in the region. There was evidence, as 2016 dawned, of a new scramble for revived influence in the RedMed by (to a degree) the United States, by (absolutely) the People's Republic of China (PRC), by the Russian Federation, and, peripherally, by other out-of-region powers: the European Union, Japan, India, and so on.

In the meantime, however, the correlation of factors is more in the hands of the regional states than it has been for centuries. How they manage affairs will determine the extent and manner in which external powers will re-enter the region. This is a dynamic and uncertain process. It impacts long-term framework issues in the Eastern Mediterranean, the Middle East, the Horn of Africa, and East Africa.

What is at stake for the global community is the stability of the critical sea lanes linking the Euro-Atlantic Mediterranean with the Indo-Pacific markets. More than that, it is part of the process which brings economic power down to the East and South of the Mediterranean, making the Mediterranean Basin trading zone more of a reality, and a reality linked to the sub-Saharan African region. It brings the growing Indian Ocean trading zone into greater interaction with the Mediterranean/European marketplace.

There are so many other factors, involving, for example: the viability (and protection) of Chinese logistics to and from the Middle East, Africa, and Europe; the ability of Russia to finally gain a stable north-south trade access (hitherto constrained by Turkey and the Northern Tier states); the structural integrity of major states in the region, such as Saudi Arabia, Turkey, and Yemen; and much more.

Traditionally significant Indian Ocean global trading states, such as Australia and India, find themselves in new territory, knowing that their "home waters" can no longer be taken for granted.

This book began its life as an intense, one-day briefing on Capitol Hill, in Washington, DC, on October 30, 2015, designed to start debate on the importance of the transformative processes underway in the RedMed. The event was undertaken by the International Strategic Studies Association (ISSA); the Gusau Institute (GI), of Kaduna, Nigeria; and the Water Initiative for Africa (WIA). Each brought separate perspectives to the process, and those perspectives form the core of this book.

The event, which brought together a select audience of officials from more than a dozen countries, was facilitated by the provision of a superb conference facility provided by Steve Ryan, a partner at the law firm of McDermott, Will & Emery, in Washington, DC. But what became clear from that day of intense discussion was that we had only begun to scratch the surface of a profound process of strategic transformation which centered on the RedMed arena.

The International Strategic Studies Association, and its associated Global Information System and *Defense & Foreign Affairs* publications, have, since 1972, focused heavily on the RedMed region, but it was significant that the new Gusau Institute — founded by the former Nigerian Defence Minister, former Army Chief of Staff, former National Security Advisor, Lt.-Gen. (rtd.) Aliyu Gusau Mohammed — made this conference its first major policy event staged outside the African Continent. As well, Prince Ermias Sahle-Selassie Haile-Selassie's Water Initiative for

Africa used the event as its first major foray into elevating discussions of water and waterways to the grand strategic arena. This was critical, given the revived significance of debate about future options for the Nile River as well as with regard to Ethiopia's revived position as a *de facto* major Red Sea power.

This book stresses that the future of the RedMed is vitally linked to the actions of near-powers Turkey and Iran (as well as the UAE, Qatar, and Oman), and on the global power factors which impact those states, whether from the US, Russia, the European Union, or the People's Republic of China. India, Australia, Japan are all party to the region's opportunities and security which affect their own.

The conference was emphatic in its conclusion that there was insufficient exchange of intelligence and perspectives within the RedMed region, and an even greater paucity of international comprehension of the region, its peoples, and the evolving nature of the strategic equation there. Moreover, the region is rarely seen within the broader global context. Lt.-Gen. Mohammed, as Chairman of the Gusau Institute, called for the creation of an independent "clearing house" for information on and between the regional states. He recommended ISSA as that independent body, with GI and the WIA participating and supporting the facilitation of a greater forum of understanding between the region and its out-of-region allies.

This book is the first stage of that process, which we hope will stimulate greater understanding and debate on the RedMed region.

We have included additional contextual material which helps place the issues of the RedMed into a broader terrain.

Gregory R. Copley
Lt.-Gen. the Hon. Aliyu Gusau Mohammed (rtd.)
Prince Ermias Sahle-Selassie Haile-Selassie

Washington, DC, and Kaduna, Nigeria: January 2016

cDermott

I

RedMed: The Global Strategic Waterways

HIH Prince Ermias Sahle-Selassie Haile-Selassie

HUMAN SOCIETIES predominantly identify the land on which they dwell as the essence of their survival and importance. However, it is the two great strands of water — the Suez/Red Sea sea lane and the Nile River — which determine the viability and global strategic importance of the Horn of Africa and adjacent areas. They represent, in this regard, the links between the Mediterranean Basin and the Indian Ocean trading basin.

It is significant that this area, the cradle of human evolution out of Africa and into the Eurasian landmass and beyond, has retained its position as the global nexus into the 21st Century. Indeed, the Nile and the Red Sea have held a greater strategic significance for all of humanity longer and to a greater degree than possibly any other great bodies of water in the world. The importance of both of these waterways today has only grown in significance as the global strategic architecture undergoes its present and anticipated round of changes.

My particular interest in water, as Patron of the Water Initiative for Africa, has perhaps made me more conscious than would otherwise have been the case about the significance of these two bodies of water. So I am happy to add my introduction to this diverse collection of thoughts, which we hope we will begin to determine ways to ensure stability and growth in the Red Sea/Suez and Horn of Africa so that we may facilitate global cooperation, rather than competition.

We are at a decisive phase in this, as wc see, for the first time in perhaps 500 years, the relative absence of a truly dominant superpower presence in the region, and yet we are at a time when the maritime traffic through the Red Sea/Suez may increase substantially, and when the populations, wealth, and capabilities of the key Nile littoral states are, again in relative terms, likely to increase dramatically.

It is probable, however, that we are only witnessing a brief interregnum in major power engagement in the region, with the real prospect that competition between out-of-region powers for influence could rise substantially in the near future. This holds the promise of both danger and opportunity for the regional states, and for the stability of the region as a conduit for global trade.

For the first time, the Indian Ocean region itself is the dynamic region of global trade, largely because of the rôles of the strategic waterways in the corners of the Indian Ocean, and particularly the Red Sea/Suez and Hormuz Strait in the north-west and the Malacca and ASEAN straits in the East. It is the area of *strategic movement*, or strategic dynamism, linking the Atlantic and Pacific oceans. What happens in the Indian Ocean and its key straits ultimately determines much of the wealth of the world.

We have also to be conscious of the fact that not only are the global traders dependent on the stability of the Red Sea/Suez sea lanes, but so, too, are some 48 littoral states of the Indian Ocean basin, as well as the littoral states of the Mediterranean. What passes through the Red Sea/Suez, in many instances, also passes through the Mediterranean.

It is fitting, then, that this study should see the Suez/Red Sea/Horn of Africa as the hub of affairs which are vital to the emerging global balance, the new world strategic architecture which is starting to take shape. The stability of this hub is critical.

We were blessed, in many ways, by the vision of my late grandfather, His Imperial Majesty Emperor Haile Selassie I, to create a major regional organization, the Organization for African Unity, in Addis Ababa. It has evolved into the African Union, of course, but it retains its unique window onto the Indian Ocean and Eurasia, as well as into the Middle East and Maghreb. We are seeing now the revived significance of Addis in the successful, if delicate, accord signed on August 17, 2015, between warring factions in South Sudan. This may yet presage an opportunity to develop White Nile water projects, possibly even including the diversion of Congo River waters into the greater Nile flow. We are now witnessing the start of cautious, but extremely positive, overtures by the Egyptian Government on Nile water use. We also see the good offices of the Egyptian Coptic Pope, His Holiness Tawadros II, to engage in society-to-society dialog to ensure that we enter a new era of cooperation on the utilization of Blue Nile waters.

It is my hope that my own Water Initiative for Africa can also serve in this building of social understandings between the people of the region, facilitating grassroots-level support for mutually beneficial water outcomes.

What we are witnessing is the possibility that — by sound and careful management — the greater Horn of Africa region could become an area in which aridity can be banished through human cooperation and ingenuity. We are also witnessing the prospect that the fortunes of the area of the Arabian Peninsula through to the Eastern Mediterranean could also be transformed through human ingenuity in water technologies, coupled with relatively abundant energy.

We should never forget, too, that the eternal clan conflicts in Somalia — often today cloaked in, or suborned by, religious zealousness — have always been about which clans controlled which wells. Those of us from the area understand the fundamental reality that water is the vital underpinning of wealth and power.

It would not be inappropriate also to consider the extent to which the Arabian littoral of the Red Sea is equally strategically delicate because the fundamental issue of water has yet to be adequately addressed. Moreover, it is probably appropriate that I should address this issue about the underlying significance of water — rather than oil — as the liquid which underpins survival and stability on both sides of the Red Sea, given that the ancient Solomonic kingdom of my ancestors once straddled both sides of the Sea under Queen Makeda of Saba, a thousand years before the Common Era.

We have many challenges yet to resolve to stabilize the Suez/Red Sea/Horn region, including, of course, those states on the Arabian Peninsula. The stability of the regional states — Egypt, Israel, Jordan, Saudi Arabia, Yemen, Sudan, Eritrea, Ethiopia, Djibouti, Somaliland, and Somalia — is critical to the entire European Union, the security of which is so dependent on its southern littorals and the trade which comes via the Red Sea. So, too, are the stability and wealth of the Indian Ocean and ASEAN powers, plus East Asia, critically dependent on the stability of the greater Red Sea which we are addressing today. And on the prosperity of these regions — Eurasia, Asia, and the Middle East — do the fortunes of Africa and the Americas depend.

It is not my task here to dwell on all the details, nor specify the answers. But I have chosen the theme of water as my mission for Africa and the greater region, so it is appropriate that I remind us all that it is the waterways which bind us, and it is this region we discuss in this book which — although it has been long neglected as the nexus of the global strategic balance — holds much of the hope for the future, if we can but master it.

I have mentioned how the world's fortunes turn in large measure around the stability and freedom of navigation of the Red Sea and Suez, and on a unique new prospect to enhance the volume and use of Nile waters. But it is equally true that the changing strategic fortunes of Israel, Jordan, Saudi Arabia, and Yemen — and other societies on the Persian Gulf and Arabian Sea — can be favorably altered by adapting new technologies to deliver desalinated ocean waters to the hinterlands. Much of this technology has been available for some time, and we have seen how the strategic fortune of Israel has already been improved by the significant adoption of an old technology — reverse osmosis — to make Israel water independent. This will be further enhanced with the availability of cheap local gas to energize the desalination, but it could be enormously enhanced by even more efficient techniques to create pure drinking water now being considered, including some developed with the aid of the Water Initiative for Africa.

Jordan is reportedly considering the use of nuclear power to enable unlimited seawater desalination from the Gulf of Aqaba. Saudi Arabia has long considered a wide variety of desalination endeavors, but it will now — in the face of a restive population and a finite life for its great oil reserves — need to consider that the best use of its energy could be to support water abundance to build the Arabian Peninsula into a green and food self-sufficient land. It could transform Saudi Arabia from an oil-dominated state to a more balanced society, and one, therefore, in which more of its society becomes occupied by productive labors aimed at a balanced agricultural, manufacturing, and service economy.

The underpinning of stability and prosperity in the region is a complex and long-term view of all aspects of water: for navigation, agriculture, and the transformation of human society. If we look today at the negotiations over the appropriately-named Grand Ethiopian Renaissance Dam, a project estimated to cost some 15 percent of Ethiopia's gross domestic product, we can see some of the prospects. It is being built near the border with Sudan, for the generation of hydro-electrical power, much of which will provide low-cost energy to Sudan. But it is built at such an altitude that it will hold waters which will be less subject to evaporation than those in the lower-altitude, higher-temperature region of Lake Nasser, lower down the Nile at the Aswan Dam. The result will be that billions of cubic meters of Blue Nile waters will be saved from evaporation each year, to be released into the system, ultimately swelling the volume of water available to Egypt and Sudan.

Of course there remain many aspects of this major project to be resolved, and national interests taken into account, just as there were when the Aswan High Dam was being built in the 1960s, and before it was inaugurated in 1971.

What we are seeing now, however, is a very positive evolution of diplomacy on the Nile. Egyptian President Abdul Fatah al-Sisi on August 25, 2015, through Egyptian Minister of Water Resources and Irrigation, Hossam El Din Maghazi, announced the signing of a cooperation agreement with Pope Tawadros II, head of the Coptic Orthodox Church under which His Holiness the Pope would assist in the bilateral discussions with Ethiopia on the Renaissance Dam project. This step, and others undertaken by Pres. Sisi, have shown a remarkable commitment to achieving a solution to water issues based on mutual needs.

In other areas relating to the Nile, the possible diversion of some of the headwaters of the Congo River to the White Nile, in South Sudan, in a great future engineering project — of which some of the work has already been done — could add even more to the Nile flow, with the prospect of transforming both Sudan and Egypt into great agricultural greenlands. Again, the Egyptian Government has a strong engagement in this process, and thus has a strong interest in seeing stability in South Sudan and Sudan.

All of these possibilities are open in this pivotal region, stabilizing not only the controlling lands of the Suez/Red Sea sea lane, but also creating one of the world's great new emerging marketplaces. This is a strategic vision worthy of consideration. We know that there are great negotiating hurdles to overcome; indeed, these may be greater than the engineering challenges.

But the rewards are worth the struggle, if first we can define the overarching vision.

As well, let me say in conclusion, we should be aware of the lessons in water usage, water diplomacy, and hydro- and energy-technology which we can translate from the Nile River Basin and the related Red Sea states to other water-challenged areas of Africa and beyond. The Niger Basin and Congo River Basin — and other water systems — have challenges which can benefit from the framework we develop from this point forward in the Nile and Red Sea regions. Indeed, these are lessons for other areas in the world, such as the Indus River Basin, which is very much part of the geostrategic framework linked to the Nile/Red Sea/Horn region we are discussing in this book.

McDermo

II

The Global Nexus at the Suez, Red Sea, and Horn of Africa

Why the RedMed Region is Becoming Increasingly Central to the Emerging Global Strategic Framework

Gregory R. Copley

We are now in a fluid, multipolar world, so it is no coincidence that the great global highway from the Mediterranean to the Indian Ocean is of growing interest and competition. It is also an area of real strategic movement in its own right.

Historical trends, including the rise of the Chinese and Indian economies, are combining once again to make this nexus a dynamic phase of the evolving global strategic architecture.

There are so many complexities regarding this region that this study alone cannot do justice to them.

It is not, of course, just the US and the People's Republic of China (PRC) which are focused — as though their future depends on it — on the Mediterranean-Suez-Red Sea SLOC (sea line of communication).

Apart from the littoral states themselves, Russia is emerging as a significant factor in the Eastern Mediterranean. On the Southern side of the Suez Canal, many other states are vitally engaged in the stability and security of this sea lane and its littorals, from Japan and India to Turkey and the European Union states, and from Iran and Pakistan to Australia, and so on.

The prospect now exists for the Suez-Red Sea-Horn region to gradually emerge as a major, integrated economic trading zone; perhaps the next major global marketplace, linking Africa, the Middle East, and the Mediterranean.

As a convenient shorthand, I'm calling this integrated strategic theater the "RedMed". It consists of Egypt, Israel, Jordan, Saudi Arabia, Ethiopia, Yemen, Sudan, Eritrea, Djibouti, Somaliland, and Somalia.

The littoral states of the RedMed have a population of some 316-million people, and a combined 2014 GDP of at least $1.5-trillion.[1] They interact, of course, with neighbors in Africa, Europe, and the Indian Ocean basin economies.

The 48 Indian Ocean states — which include the RedMed group — contain more than a third of the world population.

Major structural change in human society began with the end of the Cold War. In the Middle East and Africa, we are now seeing a reversion to more historical and basic human groups: clans; linguistic, cultural, or religious groupings; some ancient civilizational groups, such as Persia, Egypt, Ethiopia, Israel, and so on.

These re-emerging, or reviving, divisions demonstrate

1 Egypt: $286.54-billion; 88,487,396 pop.; Israel: $304.23-billion; 8,049,314 pop.; Jordan: $35.83-billion; 8,117,564 pop.; Saudi Arabia: $746.25-billion (2014); 27,752,316 pop. (2014); Yemen: $27.1-billion (2014); 26,737,317 pop. (2014); Sudan: $73.82-billion (2014); 36,108,853 pop. (2014); Eritrea: $3.858-billion (2014); 6,527,689 pop. (2014); Djibouti: $1.589-billion (2014); Ethiopia (counted here as a *de facto* Red Sea littoral state): $54.8-billion (2014); 99,465,819 pop. (2014); Somalia: $2.372-billion (2010 est.); 10,616,380 pop. (2014); Somaliland: $1.4-billion (2012 est.); 3,500,000 pop. (2008). *Source:* Global Information System.

that there are durable strands in human identity which survive thousands of years.

These factors are evident, for example, in some of the recent changes such as:

- The break-up of Ethiopia with the separation of Eritrea;
- The break-up of Sudan to create a separate South Sudan;
- The *de facto* break-up of Yemen;
- The *de facto* break-up of Iraq;
- The breaking-up of Syria;
- The *de facto* break-up of Libya; and
- The return to separate status of what had been the Italian and British Somalilands into Somalia and Somaliland, and the concurrent fracturing of the rump Somalian state.
- And there is now speculation as to the possible break-up, or breakdown, in the coming decade or so of Turkey and Saudi Arabia, two of the most significant economies in the region.

Yet this is only the tip of the iceberg; the start of a dynamic century. Neighboring Europe is, itself, undergoing massive population and cultural/political transformation, and at least two European Union member states still face significant separatist tendencies: the United Kingdom and Spain. The EU as a geopolitical entity is also, itself, undergoing unplanned transformation as we see separatist groups emerging, the start of the re-assertion of nationalism, and population composition altering the nature of societies.

Strategic transformation is now occurring on a global scale. This will largely be as a result of what seems to be a gradual return to a bipolar world, but that may or may not occur as anticipated. It almost certainly will not mirror the bipolar world of the last half of the 20th Century.

The RedMed region is at least as critical, at least as complex, and less stable — in positive as well as negative terms — as the concurrent South China Sea/ASEAN[2] theater. These regions are very much interlinked, because the Indian Ocean is, for much of global trade, the transit space, not the destination.

In other words, much of the trade which enters or leaves the Indian Ocean through the Red Sea/Suez sea lane or the Strait of Hormuz, also enters or leaves it through the ASEAN waterways and the South China Sea. But that, too, is changing.

The recent expansion of the Suez Canal also means that the Red Sea/Suez SLOC will become even more significant because of its preferred status and preferred economics over the Cape of Good Hope sea route. It is arguably the most critical SLOC in the world.

The nexus within this "Gibraltar to Socotra strategic space", then, is the RedMed theater. This is the pivotal geographic junction of global trade and energy. It is, as well, a key component in the question of refugee flows to Europe. But for the first time in centuries, the regional powers hold considerable dominion over their own fortunes.

To gauge the impact of the Red Sea nexus, we can look back at the post-World War II pattern, at the true cost to the global economy of the closures of the Suez Canal in 1956-57, and between 1967 and 1975, and even during the brief interruption to shipping caused by the mining of the Red Sea by the Libyan minelayer, *Ghat*, in July 1984. Today, the Red Sea/Suez Sea Lane is substantially more important.

But there is much more to consider than the region's effect on the world. We must also consider the world's effect

2 ASEAN: Association of South-East Asian Nations, consisting of Indonesia, Malaysia, Philippines, Singapore, Thailand, Brunei Darussalam, Vietnam, Laos, Myanmar, and Cambodia.

on the region. Global transformations and fluctuations affect the fortunes of the region, as we saw with the decline in Suez Canal revenues in 2008-09 with the global economic crisis.

To start with an overview of the possible evolution of the region, we need to consider that several key states may not exist in their present or 20th Century form or relative power within a decade: Libya, Turkey, Saudi Arabia, Syria, Lebanon, Iraq, Yemen, Sudan, and Somalia.

There are varying degrees and forms to this change, or restructuring.

If, hypothetically, Saudi Arabia was to fracture, the prospect is that Qatar, and possibly the UAE, or even Jordan and Yemen (or some part of what is now Yemen), could expand to reclaim some of their former clan areas. And what happens in the Arabian Peninsula affects what happens on the Horn of Africa, and *vice-versa.* This has been the case since the cross-Red Sea Kingdom of Queen Makeda of Saba, some three millennia ago.

On the other hand, if Saudi Arabia could survive the present downward spiral of its finite resources and low energy prices, it would be because of a strong alliance and trading structure with, for example, Egypt and possibly Israel, and potentially Ethiopia. And it would be, among other reasons, because the Saudi Arabian economic model transformed from its present dependence on oil, gas, and investment returns, through new water purification and distribution means, into a more balanced economy, and one which was able to limit, rather than grow, its population.

Of pivotal importance because of its regional reach would be the possible break-up or re-shaping over the coming decade of Turkey, which has historically and currently been active in the Maghreb, Red Sea, and Horn. This, along with possible changes on the Arabian Peninsula, would substantially alter the way in which out-of-region powers (such as Russia, the PRC, and the US) are channeled in their engagement in the area.

It would probably lead to the return of Egypt, Iran, and Ethiopia as anchoring poles of the greater Middle East, because they represent actual historical civilizations.

Iran and Egypt each have substantial, but not overwhelming, economies; indeed, their economies are not substantially larger than that of Israel, and Israel's *per capita* economic wealth dramatically exceeds that of either Egypt or Iran. Ethiopia's economy, however, is nascent, but growing, and could grow even more rapidly.

What is critical about the revival of Egypt, Iran, Ethiopia, and Israel is that they represent cohesive civilizational structures which can command a sense of national unity and productivity. Most importantly, they all function — for the first time in generations — as fully sovereign entities, making strategic decisions based on their own perceptions of national requirements.

Egypt, Iran, and Israel are making strategic decisions today largely without reference to other powers. Ethiopia is once again re-emerging in this context after four decades of difficult internal preoccupation, much of it governed by Cold War politics. That does not mean that these civilizations have no allies; nor that they disregard external influences. But they have determined their own strategic goals and are moving toward them.

Within this framework, while we are seeing already the significant restructuring of Egypt, it is worth noting that we should start to see the restructuring of Iran begin to become visible within a few years. And similarly with Ethiopia.

What is emerging, then, in this broad arena of the

RedMed are two overlapping *blocs* of states — North and South of the Suez, in essence — which share many vital and direct interests.

In the North and Western *bloc*, Egypt, Israel, Jordan represent a core. Egypt and Israel link the new Eastern Mediterranean energy resources with Cyprus and Greece.

Lebanon, Syria, and Turkey could have shared interests with this core *bloc* of Egypt, Israel, Cyprus, and Greece, but they are at present peripheral to the *bloc* because of political issues.

The Southern and Eastern *bloc* states of Saudi Arabia, Ethiopia, and Djibouti share vital interests with the North-Western *bloc* of Egypt, Israel, and Jordan. There are other regional states which should also share these interests but which, in the near term, may not be part of the equation because of political differences: Sudan and South Sudan, Eritrea, Yemen, Somaliland, and Somalia.

Oman, although outside the Red Sea, nonetheless also has a significant stake in the stability of the Red Sea littoral, and, significantly, is opposed to the Saudi-led conflict in Yemen.

Yemen, the Sudans, Eritrea, Somaliland, and Somalia have preoccupations at present which keep them partially or wholly marginalized from the core Red Sea *bloc* of Egypt, Israel, Jordan, and Saudi Arabia. And even Saudi Arabia is itself gravely distracted at present, because of economic issues and its debilitating involvement in the conflicts which threaten to break up Yemen and Syria.

Some key states just outside the region are also involved in either helping or destabilizing the Red Sea region: Iran, Qatar, Kuwait, the United Arab Emirates, and Kenya.

And a number of major external powers — the US, France (and the EU generally), the People's Republic of China, Turkey, India, Pakistan, Japan, Australia, and Russia

— are also engaged in the protection or projection of their own interests into the Red Sea / Horn of Africa region.

Indeed, the 30-nation[3] Combined Maritime Force (CMF), based in Bahrain, in the Persian Gulf, has enabled a significant number of maritime states to express their involvement in the security of the region, through participation in the combined task forces, particularly CTF-150 on counter-terrorism and maritime security, and CTF-151 on counter-piracy. These have been particularly helpful, for example, in reducing Somalia-based piracy against commercial shipping.

Engagement in the anti-piracy work enabled Japan and the PRC — neither of which is in the CMF — to significantly expand their long-term maritime presence in the north-western Indian Ocean.

But it is Djibouti which is a key pivot point in international engagement in the region. The US, France, and Japan already hold secure basing in Djibouti. The PRC has secured a long-term agreement to base naval and other forces in Obock, on the northern shore of Djibouti's Gulf of Tadjoura. Djibouti was reportedly seeking a $100-million a year rental. The Obock site would physically separate it from the French, US, and Japanese basing.[4]

A profound divergence in the Djibouti relationship with Saudi Arabia and the United Arab Emirates emerged in April 2015, leading to a Saudi-UAE move to withdraw Gulf Cooperation Council (GCC) troops from their basing in

3 Australia, Bahrain, Belgium, Canada, Denmark, France, Germany, Greece, Italy, Japan, Jordan, Republic of Korea, Kuwait, Malaysia, the Netherlands, New Zealand, Norway, Pakistan, The Philippines, Portugal, Saudi Arabia, Seychelles, Singapore, Spain, Thailand, Turkey, UAE, United Kingdom, United States, and Yemen. The CMF is commanded by a US vice-admiral who is concurrently Commander of the US CENTCOM naval forces, and Commander, US Fifth Fleet; with a UK Royal Navy commodore as deputy commander. Command of the three tasks forces rotate among member states.

4 See, Copley, Gregory R.: "The PRC, With its Djibouti Move, Commits Globally", in *Defense & Foreign Affairs Special Analysis*, June 11, 2015.

Djibouti — from whence operations were being conducted into Yemen — leading to the investment of considerable Saudi and UAE efforts to build basing relationships in Eritrea and (as of late November 2015) Somaliland (discussed in more detail in Chapter VIII).

The Djibouti-PRC basing relationship has, as noted in a report in August 2015, put Djibouti strategically "in play".[5] It is "in play" because the US Government, if it wishes to sustain any dominance in the region, must attempt to constrain, contain, or balance the PRC projection.

The PRC relationship with Djibouti and Ethiopia is already well-established, and the PRC recently completed the new Djibouti-Dire Dawa-Addis Ababa rail link which is of profound strategic significance to Ethiopia and Djibouti. It is also vital to China in its hope to dominate logistical lines for resources leaving Africa, and Chinese products entering the Continent. Plans exists for the rail link to be extended from Addis Ababa to Juba, the South Sudan capital, and then on to Lamu Port, in northern Kenya. This provides a valuable expansion of regional infrastructure.

[The signing of an agreement in October 2015 to build a 550km long, $1.4-billion, 20-inch fuel pipeline from Djibouti to Awash, in Central Ethiopia, is also a significant sign of the emerging infrastructure. That project will be complete by 2018.]

Resolution of the civil war in South Sudan is critical for the Lamu Port-Southern Sudan-Ethiopia Transport Corridor project[6] (known as LAPSSET), and thus to the increasingly integrated economies of Ethiopia and Djibouti. The August 2015 "peace accord" between the factions of South

5 Copley, Gregory R.: "Djibouti Takes Center Stage as Horn, Arabian Peninsula, and the Muddied Waters of the Red Sea Are 'In Play'", in *Defense & Foreign Affairs Special Analysis*, August 4, 2015.

6 See "A New Watershed Conflict Emerges in Africa as the PRC Battles to Push Resource Flow to the Indian Ocean", in *Defense & Foreign Affairs Special Analysis*, October 15, 2013.

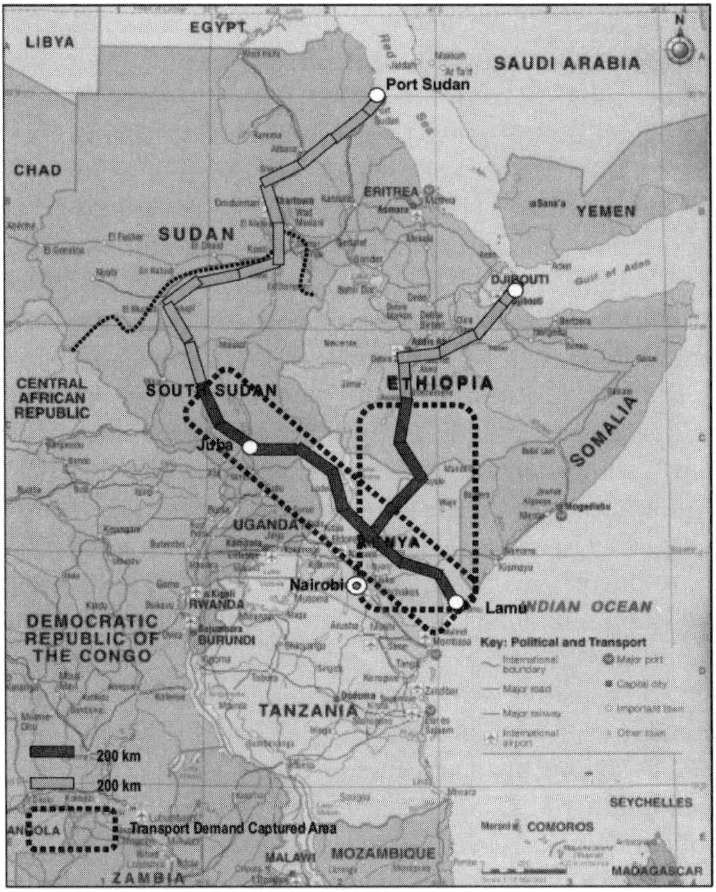

The LAPSSET corridor linking Djibouti through Ethiopia and South Sudan to Lamu Port in Kenya, but with additional links planned up through to Port Sudan.

Sudan may not, however, have resolved the civil war or the threat of ongoing hostility toward South Sudan from Sudan. China has invested heavily in terms of efforts and funds to try to stabilize the South Sudan and Sudan-South Sudan situations, as have the African Union and others.

We know, too, that, as part of this, China is negotiating

the deployment of naval assets to basing at Lamu — apart from the commercial endeavor of the port — and justifying this, as well as its Djibouti basing, as part of its ongoing anti-piracy patrol in the region.

But on the immediate question of Djibouti, we are seeing that the PRC basing arrangements in Obock has finally set off alarm bells in Washington. In a report on October 2, 2015, I noted that Djibouti was in a dynamic state; that is, in a state of flux, and suggested that we should expect it to come under massive US pressure soon.[7]

What is significant is that the US response to the growing encroachment of China in Djibouti and the region has not been met as well there as Washington would have liked.

On May 6, 2015, just as Pres. Guelleh was in the midst of negotiations with the PRC on Obock, US Secretary of State John Kerry visited the country and was portrayed as having publicly criticized the President for alleged human rights violations. This was only a minor point in what was a mission to thank Djibouti for its cooperation in the war against terrorism, but it was taken up by Djiboutians who were opposed to Pres. Guelleh. Similarly, US Pres. Barack Obama's comments in Addis Ababa on July 28, 2015, in support of gay marriage in Africa created a political backlash against the US, undoing some of the positive impact of a visit which was meant to cement the close US-Ethiopian security ties.

What, then, would be Beijing's options, if US pressure against the PRC's Obock basing was successful (unlikely), knowing that the PRC must have a stake in the control of the Red Sea? Perhaps to move toward the use of Berbera, the great naval basing option in Somaliland? That would imply some additional steps by Beijing to stabilize and transform

7 Copley, Gregory R.: "All Change in the Middle East? Syria, Egypt Trend Positively; Saudi, Iran Transform", in *Defense & Foreign Affairs Special Analysis*, October 2, 2015.

the Republic of Somaliland, and to upgrade the Berbera to Ethiopia transportation links.

At this stage, if anything, the People's Republic of China is as committed as the US — perhaps even more so — to securing a naval presence which would help facilitate its transportation of vital resources out of Africa and the Middle East.

And if not Obock or Berbera, then Beijing's options would have to include the Eritrean ports of Assab or Massawa, even though the stabilization of Eritrea and the mending of Ethiopian-Eritrean links would be far more difficult and expensive than mending the problems of the now dysfunctional Administration of Somaliland in Hargeisa.

In January 2015, a PRC company contracted to undertake a $400-million expansion of the port of Massawa, to include a 70,000 ton bulk cargo terminal and a 50,000 ton multi-purpose terminal. This was the biggest commercial undertaking in Eritrea since independence. This also indicated the strength of the mineral projects starting to get underway in Eritrea.

The Chincse firm, CHEC (China Harbor Engineering Company) was also engaged in building new port infrastructure in Walvis Bay, Namibia, and in Venezuela: both ports which are integral to the PRC's emerging global naval strategy, which is tied closely to its trade ambitions and actualities.

[It is also not coincidental that China's Landbridge Group, on October 13, 2015, bought control of a 99-year lease on the strategic Australian port of Darwin, at the Eastern end of the Indian Ocean, and at the southern end of the ASEAN straits up into the South China Sea. A Chinese firm, Hutchison Whampoa, in 1996 acquired control of the Panama Canal operations, and the expansion of that Canal now underway is seen as critical to the expansion of Chinese trade with Latin America. The irony is that Panama has diplomatic relations with the Republic of China (Taiwan), rather than the PRC. There is little doubt that Beijing sees the vital nature of acquiring the key maritime assets at the choke-points, and has moved through economic means to acquire them, rather than starting through military projection.]

For Beijing, the flag follows trade, rather than the other way around.

In the cases of Djibouti, Eritrea, and Ethiopia, it is already apparent that Beijing has interests in maintaining the stability and prosperity of the Red Sea region, without even yet addressing its interests in Saudi Arabia, Sudan, and Egypt.

The bottom line is that the PRC is not going to leave the region: not the north-western Indian Ocean nor the north-eastern end of it.

Its massive and vital commitments to building and dominating the port infrastructures — and transportation links — at Lamu, in northern Kenya, and Gwadar, in Pakistani Baluchistan, absolutely require that centerpiece which the Djibouti port facilities would represent. Beijing is investing heavily in developing the Gwadar to China link by expanding the Karakoram Highway road and pipeline links because it is absolutely committed to its strategic network which includes Djibouti. But it does, with Massawa, have a fallback position already underway, if necessary.

[As an aside, India is also making moves into the Western Indian Ocean choke-points. From October 26 to 29, 2015, the Indian Government hosted more than 40 African heads-of-state at the third India-Africa Forum Summit, this time being the first that all members of the African Union were invited to attend.]

So whatever steps the US takes in Djibouti — and it is already taking steps there — it would need to consider the downstream impact on Beijing's decisionmaking. The US is heavily invested in Djibouti, and has to weigh its options.

The US combined military presence at Camp Lemonnier, Djibouti, which hosts some 4,200 US service personnel, serves as a vital support for the US and allied maritime presence on the Bab el-Mandeb. It also linked, until September 2015, with the intelligence, surveillance, and reconnaissance (ISR) capabilities of the US 17th Air Force, based at the 2,800 meter Ethiopian airfield at Arba Minch, in the country's south, to a large degree represented in the form of RQ-9 *Reaper* unmanned aerial vehicle (UAV) operations. However, the US suddenly closed its operation at Arba Minch in late 2015.

The Djibouti and Arba Minch operations linked with the US facilities at Diego Garcia, in the British Indian Ocean Territories, and the US air and naval support facilities in Bahrain, and al-Udeid Air Force base, which hosts the US Central Command and US Air Force Central Command headquarters near Doha, Qatar. And, of course, with other less formal basing arrangements in the region. US operations in Djibouti are, of course, key to its surveillance operations over Yemen, and therefore critical in providing targeting information to the Saudi coalition fighting the Zaidi forces there, and in monitoring Iranian maritime traffic in the area.

All segments of this greater region are interactive. So it is clear that, in the lower Red Sea region, actions on one side of the Red Sea affect conditions on the other side.

For a number of years, young Eritrean men fled their country to escape military service. Those who could not find safe-haven in Ethiopia, or Sudan, or Djibouti, took the short sea-voyage to Yemen. Now, the war in Yemen has forced the exodus of military-age men from Eritrea to find their way across to North Africa, and thence to Europe. This is no small surge of people. Eritrea is breaking down, particularly with the decline in support from Egypt, Israel, Libya, and other former supporters. Egypt no longer wishes to pursue the Mubarak- and Morsi-era confrontation against Ethiopia; Cairo now prefers diplomatic solutions in its discussions with Ethiopia, and no longer supports Eritrea's covert war against Ethiopia.

Libya, without Qadhafi, also no longer funds Eritrean-run guerilla operations into Ethiopia, which it had run through Somalia and Somaliland, as well as directly across the Ethiopian borders.

The aggressive opportunism of Eritrean Pres. Isayas Afewerke seems close to an implosion, but the prospect exists that the 69-year-old former revolutionary will lash out in a final attempt to reassert some regional relevance and to preserve his Government. Certainly, he had, until the recent mineral and port deals, done little to build an economic base in his state.

The ongoing collapse of Eritrea has been compounded by the decline in support for Eritrea from some of its foreign sponsors, particularly Egypt and Libya, which were anxious — during earlier governments — to dominate the Red Sea and to ensure that Ethiopia, a former Red Sea power, was unable to hold sway over the mouth of the Red Sea. Now, Libya is in disarray, and the Egyptian Government of Pres. Abdul Fatah al-Sisi is committed to a strong working relationship with Ethiopia on a range of issues, particularly the Nile water usage.

The linkages between the Egyptian and Ethiopian societies — as well as their governments — are building, despite as-yet unresolved governmental differences on the modalities of such things as Nile water use. But the case for cooperation is now seen as more appealing than the case for confrontation.

So Eritrea has lost many of its sponsors, other than Qatar — which supports the Muslim Brotherhood and is therefore seeking ways to maneuver against Egypt — and now Saudi Arabia and the UAE, and Turkey, which are seeking leverage in the Red Sea/Horn region. Iran's options in the Red Sea have been confronted by the moves in 2015 by Riyadh and the UAE into Eritrea.

Pres. Isayas, in poor health in recent years, has, as part of his posture, hinted at a reunion or confederation with Ethiopia, of which Eritrea was historically a significant part. Ethiopia could not consider this while Isayas remained at Eritrea's helm.

But if the collapse of Eritrea is of concern, then equally, too, is the war in Yemen, now underway as an international conflict for nine months, spearheaded by Saudi Arabia and a coalition of states. What is of concern in the Yemen war is that a complete success by Saudi Arabia — aimed at restoring the Government of Pres. Abd al-Rab Mansour al-Hadi — might be difficult, given that the war has now re-opened sectarian as well as geopolitical divides in the country.

An early success in the war against the Houthi-led forces, which still control much of the north, including the capital, Sana'a, was, by early 2016, critical to Saudi Arabia if the Kingdom was not to suffer significant financial consequences at a time when its economy was challenged.

The question is whether the Saudi coalition could, indeed, succeed in being able to restore a sustainable government of the entire Yemen, or whether any patchwork solution would require such "deals with the devil" as to render any follow-on government in Yemen either paralyzed or weak.

Extensive areas in southern Yemen, extending over toward the Hadhramaut, are already under the control or influence of radical *jihadist* groups which are part of, or allied with, *al-Qaida*.[8] It has been argued that the Houthi Zaidi forces were, as a Shi'a sect, always allied with and dependent on Iranian support, but this seems to be an overstatement of the position. It is more likely that the conflict — which expanded out of control, and brought in Saudi support for Pres. Hadi — may have opened up an opportunity for Iran to inject itself further into the conflict, especially given Iran's support for the earlier declaration of the so-called "Islamic Republic of Eastern Arabia" in the Shi'a areas of Saudi Arabia in May 2009.

And Iran's long-term engagement in the Red Sea and Horn region is something which pre-dates the clerical Government which took office in 1979. But even now, before Iran has recovered economically from the effects of years of sanctions, and while its government is in many respects dysfunctional, it still manages to control much of the strategic dynamic in the region. It has been able to achieve this with a defense budget, in 2014, of only around $8-billion. Saudi Arabia, before it ramped up defense spending to deal with the Yemen crisis, had a 2014 defense budget of around $80-billion.

What is significant in the Yemen conflict is that Qatar and Saudi Arabia are financing the use of Sudanese (and Eritrean) troops to fight in Yemen to support the anti-Houthi/anti-Iranian cause. Entire units of the Sudanese Army are being rented, on a mercenary basis, for this cause. And yet, on the other side the coin in Sudan, the Government of Pres. Umar Hasan Ahmad al-Bashir is working

8 In Abyan Governorate of South Yemen, adjacent to Aden, the situation is complicated by the existence of armed groups led by "popular committees" representing *al-Hirak* [Southern Movement, advocating secession from the Yemen Republic], *al-Islah* [Yemeni Congregation for Reform; Wahhabists linked to Muslim Brothers/*Ikhwan*], *salafis*, and Sunnis. *Al-Islah*, for example, was known to have received Saudi funding in the past, but its *Ikhwan* links may have made it less palatable to Riyadh in the past year or two (or the group may have opted to align with Saudi Arabia and abandon the *Ikhwan*).

equally as fervently with the Iranian Government to support Iranian objectives in the region.

We have not, in this brief overview, had the opportunity to discuss in depth the other littoral states of the Suez/Red Sea/Horn, namely Sudan, Israel, and Jordan. They have, of course, key rôles.

What is clear is that the two significant regional naval powers, Egypt and Israel, will increase their capability, of necessity, into the Red Sea and the Indian Ocean. We will discuss Egypt's transforming security capability later.

Israel's expanding submarine fleet — which will comprise six *Dolphin*-class, or modified Type 212 boats, built by HDW in Germany — will give that navy an expanded capability in the Red Sea, possibly supported in the future more visibly by new Israeli surface combatant ships.

In all of this it is conceivable to see a positive, but challenging period ahead for the players in this global nexus. The biggest negatives, of course, are the conflicts and competitions of the region. Those include:

- Issues between Egypt and Ethiopia on the resolution of Nile and Red Sea issues because, despite significant differences in negotiating styles, both states have far more to gain from cooperation than confrontation;
- Problems between South Sudan and Sudan, which is a less stable and predictable situation, but where cooperation could be the link which integrates the economies of East and Central Africa with Egypt and the Horn. On the other hand, the growing radicalization of Sudan has marked it as a major resource of *jihadist* support for fighters targeting such diverse states as Egypt and Nigeria;
- Conflict between Saudi Arabia and the Yemeni Zaidis, where there is enormous long-term potential for damage unless a resolution could be achieved within, say, a year, and even then there will be serious legacy prob-

lems and almost certainly polarized societies;
- Problems between Ethiopia and Eritrea, which may only be resolved when Eritrean Pres. Isayas Afewerke departs the scene; and
- Problems between the warring factions in Somalia, which show no sign of an early resolution, although, as with the Puntland region, the combatants have been showing signs of fatigue, despite being one of the fountainheads of *jihadist* support for African militant groups. And problems between Somalia and Somaliland.

The biggest positives are based around the prospect of a substantially and rapidly growing Egyptian economy, linked with the regional energy basin in the Eastern Mediterranean. That alone brings Egypt and Israel into a common economic framework with Cyprus and Greece, and gives the ability to help alleviate economic challenges in Jordan, for example. It may ultimately also have a moderating effect on relations between the Palestinian Authority and Israel. Even, perhaps in the coming years, re-opening trading relations between Iran and the RedMed states.

Within this, too, is the fact that out-of-region powers are competing for influence in the region. While this poses some dangers, it also has already demonstrated that investment into the region has increased and shows signs of increasing still further.

It is unlikely that the region will become less militarized, although Ethiopia, with its budget challenged by its expenditure on the Grand Ethiopian Renaissance Dam, may be unable to expand its military as much as it would like. Saudi Arabia seems set to increase defense spending, particularly on military operations, so its budget pressures may induce it to spend less on religious proselytization issues, which may actually help stabilize Saudi relations with Ethiopia and, indeed, such states as Pakistan.

But the RedMed will be a region in which, for the first time, local players will be increasingly capable and dynamic. The trade patterns are already changing, too. The Indian Ocean basin, and particularly the East African region, is beginning to show signs of activity, because of China's demand for resources, and the PRC's need for sea lane security is pushing it to direct its trade up the East African coast and across to Gwadar, for carriage across the land bridge of the Karakoram Highway into China.

In all of this, then, the security of the Suez Canal is of constant concern, to an even greater degree than historically. But the value of the Suez assumes that Europe and China remain as substantial markets for the foreseeable future, requiring the full capacity of the Canal to facilitate trade. The Canal, of course, competes with the overland "Great Silk Route", so the Suez Canal will require greater trade with Europe to come from the Indian Ocean basin itself.

Perhaps the real concern is not that the RedMed is a source of instability affecting the world, but the reverse: The prospect that flattened economies and political stagnation and flat energy and resource demand in Europe and China, and perhaps in the US, could be the factors which adversely affect the RedMed.

But at least, for the first time, there is the prospect of a substantial regional marketplace in the RedMed which could grow significantly apart from the old worlds of Europe, the Americas, and Asia.

III

The Djibouti Trigger

Gregory R. Copley

SAUDI ARABIA'S WAR AGAINST THE HOUTHI in Yemen was, by early 2016, transforming more than the Arabian Peninsula. It was also having a significant impact on the African littoral of the Red Sea, which had become an important launching pad for anti-Houthi operations across the narrow waterway.[9]

The evolution of the conflict had, by the dawn of 2016, become a process which had already caused rifts between some Horn of Africa states and the Saudi-led Coalition which had been conducting Operations *Decisive Storm* (March 26-April 21, 2015) and *Restoring Hope* (April 22, 2015-present) against what were essentially the Shi'a elements in the area which had once been the Yemen Arab Republic (North Yemen).

Djibouti, which continues to host a number of foreign military powers, had also hosted a combined contingent from the Gulf Cooperation Council (GCC) (excluding Oman, which had refused to support the Saudi-led war in Yemen). But Djibouti's hosting arrangements changed when a diplomatic difference arose between Djibouti and

9 Belligerents include, on the one side: Yemen (Hadi Government); Saudi Arabia; UAE; Bahrain; Kuwait; Qatar; Egypt; Jordan; Morocco; Senegal; Sudan. Supported by Somalia (non-combat); the United States (intelligence, weapons, and blockade); and the United Kingdom (intelligence, weapons, and blockade). Opposing: the Houthi-led coalition, which includes *Ansar Allah*, the Yemen Army, Yemen Republican Guard, and the Yemen Air and Air Defense Forces, supported by Iran.

the United Arab Emirates (UAE), which resulted in the closure of the UAE Consulate in Djibouti on April 28, 2015. The closure came after what was described as "an altercation" between Col. Wahib Moussa Kalinleh, Commander of the 250-man Djibouti Air Force (*Forces aériennes djiboutiennes:* FAD), and Ali Al Shihi, the Vice-Consul of the UAE in Djibouti.

This was the formal reason of the departure of GCC troops based in Haramous, Djibouti, in early April 2015. Saudi Arabia and UAE immediately redirected their efforts to build a military base in Eritrea. The promise of Saudi and Emirati funds caused Eritrean Pres. Isayas Afewerke to cut off his links with the Houthi Shi'a leadership in Yemen.

There was more to it than that. The break followed Djibouti's unilateral cancellation of the lucrative Doraleh Container Terminal concession contract from the UAE-based company, DP World, which it then handed to a Chinese (PRC) company. UAE Pres. (and also *Emir* of Abu Dhabi) *Sheikh* Khalifa bin Zayed al-Nahyan then broke off diplomatic relations with Djibouti on May 4, 2015.

In the build-up to this, Saudi Arabia denied a visa to Djibouti Foreign Minister Mahamoud Ali Youssouf and declared him *persona non grata* in the Kingdom. Pres. Guelleh had hoped to despatch the Foreign Minister to Riyadh to persuade the Saudis to intervene in the diplomatic fallout between his country and the UAE. The visa denial was his answer; the declaration that the Foreign Minister was PNG (*persona non grata*) emphasized the brutality of the response.

That was when Djibouti turned to the US to mediate, given that the dispute with the UAE automatically extended to Saudi Arabia. US Secretary of State John Kerry — who was in Djibouti on May 6, 2015 — declined to intervene. Little wonder that Djibouti then also felt that it was compelled to move more rapidly in building relations with the PRC; this Djibouti Pres. Ismaïl Omar Guelleh did on May 9, 2015.

Meanwhile, King Salman bin 'Abd al-'Aziz al Sa'ud of Saudi Arabia gave a red carpet reception in Riyadh to Eritrean Pres. Isayas Afewerke, Pres. Guelleh's — and Ethiopia's — key strategic adversary. The King had already, on April 29, 2015, signed a contract for military and security cooperation with Eritrea. As part of this, Saudi Arabia and the UAE also offered to finance a major upgrading of Eritrea's principal airport in Asmara, and to invest in maritime and other basing options along the Eritrean coast and islands.

Pres. Isayas' close and discreet relations with Iran, which had been using Eritrean facilities to support the Houthi forces in Yemen, were being hastily wrapped-up.

By November 9, 2015, it was reported that the UAE had been negotiating to lease the Eritrean port of Assab for 30 years. This would give the UAE and Saudi Arabia a number of advantages: it would provide a base for strikes against their opponents in Yemen, while depriving Iran of the use of Eritrea to support the Yemeni Shi'a forces. The Saudi-led team had already struck deals with Sudan to hire Sudanese military units in the war against the Yemeni Shi'a, largely to ensure that Khartoum would reduce its cooperation with Iran.

The process of inserting Saudi and UAE forces into Eritrea paved the way for Saudi, Emirati, and Eritrean activities against Ethiopia. Certainly, if the Saudi coalition was strenuously opposed to the Djibouti Government, it was already striking at the heart of Ethiopia's strategic interests, given Ethiopia's growing economic links with Djibouti as its entrepôt port. Thus, the anticipated improvement in Saudi-Ethiopian relations seemed now to have become stalled.

Saudi-financed Wahhabist mosque-building in Ethiopia has been continuing apace for decades, often through the "investment" activities of half-Ethiopian Saudi *Sheikh* Mohammed Hussein Al Amoudi. As with Saudi mosque-building operations elsewhere (particularly in Pakistan, Afghanistan, and the Balkans, for example), this process had been designed to proselytize the Saudi Wahhabist sect of Islam, which has been central to the evolution of radical *jihadist* operations, including DI'ISH (Islamic State).

This has, for some time, been a point of concern in Ethiopia, which had been predominantly Christian since the Fifth Century CE. Indeed, even the Prophet Mohammed had expressly forbidden his followers to proselytize or oppress the Ethiopians after Ethiopian King Armah (Negash) — who had himself been born in Mecca — gave protection in his capital, Axum (in present-day Tigré province), to the Prophet's relatives who had been seeking safety from attacks during the time of Mohammed's campaigns on the Arabian Peninsula.

This injunction in the Islamic *hadith* appears to have been put aside by the Saudi leadership.

What is significant is that Egypt was committed to assisting Saudi Arabia and its coalition in attempting to stabilize the situation in Yemen. Indeed, this was critical for Egypt if it is to maintain security in the Suez/Red Sea sea line of communication (SLOC). However, it was also critical for Egypt to develop a stable partnership with Ethiopia, which controls the headwaters of the Blue Nile. It was for this reason that Egypt abandoned its historical support for Eritrea's covert war to destabilize Ethiopia.

Meanwhile, the UAE, in particular, began by mid-2015 using an Eritrean port for operations against Yemeni objectives. A recent UN report also claimed that 400 Eritrean troops had been deployed out of Assab to Aden to support the Saudi-led coalition. Meanwhile, other reports out of Abu Dhabi indicated that the UAE was pursuing the prospect of a lease on the Somaliland port of Berbera, something which the increasingly dysfunctional Government in Hargeisa was supporting, but in which it faced opposition from the population of Sahel province, which contains Berbera.

Other reports, from Eritrean sources, indicated that Saudi Arabia was in discussion with Eritrean officials to utilize Eritrea's covert access to the mono-ethnic communities of Ethiopia's Ogaden and Oromo regions, where Eritrea had, for decades been conducting subversive operations. Eritrean officials attempted to persuade the Saudi Government that the Ethiopian Government was conducting anti-Saudi politics and, *de facto*, destroying all pro-Saudi Islamist entities.

This has not been borne out by Ethiopian Government activities, although the Government in Addis Ababa has become increasingly concerned with external Islamist attempts to radicalize the historically loyal Ethiopian Muslim community.

McDermo
Will & Eme

IV

Controlling Crisis in the Horn of Africa and the Red Sea

Lt.-Gen. Aliyu Gusau Mohammed (rtd.)

AFRICA AND AFRICANS HAVE PAID A GREAT PRICE for the conflicts and instability on the continent and the adjoining African areas of the Red Sea since the Cold War. African forces — through the United Nations and the African Union missions — have borne a disproportionate burden of, for example, the ongoing conflict in Somalia, a conflict which had its origins more in the great power rivalry of the Cold War than in African causes.

We can say this while also acknowledging the great contributions and sacrifice in Somalia of peacekeepers from Pakistan, India, Bangladesh, the United States, and other out-of-continent forces.

So before anything else, we must also acknowledge the vital work which has been undertaken over the years in conflict resolution, conflict arbitration, and conflict prevention by the Organization for African Unity and its successor, the African Union; as well as by the Arab League, and the regional African bodies. In the case of the Horn of Africa, IGAD — the Inter-Governmental Authority on Development — and the East Africa Standby Force are the principal mechanisms for regional peace enforcement. Without their mediation and, often, intervention, we would have seen even greater conflict in the African and Red Sea regions.

Today, we must also acknowledge that the global strategic climate has changed — and continues to evolve rapidly — thereby altering the circumstances in Africa and the Middle East, particularly in the Red Sea/Horn of Africa region. At the same time, transformative technological and social trends have altered the way in which societies interact and the way wars are fought and conflicts are resolved. Regardless of the way in which global and strategic power relationships pan out over the coming decades, it is unlikely that the mechanisms of competition will resemble those of the 20th and early 21st centuries. Yet we are left in this region with conflicts in Somalia and Somaliland, Ethiopia and Eritrea, Sudan and South Sudan, Yemen, and even the downstream consequences — positive and negative — of the Sykes-Picot agreement and the Balfour Declaration on national identity in what was once the Ottoman Empire.

History is always built upon legacies.

It is necessary, then, to review how we deal with the legacy conflicts and situations we have inherited from the Cold War and post-Cold War periods, and how we cope with new pressures on borders and with new conflicts. Do our existing political, diplomatic, military, and intelligence capabilities work well for us under these circumstances? Is our use of older political-military formations and doctrine appropriate to the newer, more fluid types of conflict and diplomacy?

In the context of the Suez/Red Sea/Horn of Africa region, we need to look at a range of conflicts of varying intensities — many of them legacy conflicts from an earlier period of Cold War rivalries — which remain of concern:

- The conflict in Yemen is now an international conflict.

It affects not only the security of the Red Sea shipping lanes, but also is a factor in the refugee crises impacting both sides of the Red Sea. The result is the flow of refugees out of the region, across the North African littoral to Europe. The political structures and shape of Yemen itself grew out of Cold War rivalries, as well as from age-old tribal and cultural loyalties. External influences into Yemen have been a spur to global terrorist operations, impacting not only the West, but also inciting radicalism in sub-Saharan Africa, as we in Nigeria have learned to our peril;

- The conflict in Somalia shows little signs of abating, although we have seen some amelioration of cross Red Sea activities in relationship to, for example, piracy. This conflict, too, has aspects of its origins in the post-colonial political structures which were then subject to Cold War rivalries. We continue to see regional ramifications of this, both in the outflow of militant activity into Kenya, Ethiopia, Somaliland, and in the requirement for African and international peacekeepers in Somalia. Ethiopian Prime Minister Hailemariam Desalegn, at a UN summit on peacekeeping, on September 28, 2015, for example, committed an additional two battalions of troops to its already-extensive peacekeeping force with the African Union Mission in Somalia — AMISOM — and yet there is little hope in sight for a settlement of Somalia's problems. British Prime Minister David Cameron also announced at that September 28 summit that Britain will send troops to Somalia and South Sudan. What happens in Somalia, in terms of the new impetus of radical terrorism, flows around the world, and has, for example, linked to the *Boko Haram* crisis in Nigeria;
- The internal unrest in Eritrea, which has caused a significant refugee flow into neighboring areas, and ultimately into Europe, particularly as the refugee outflow was moving partly into Yemen are now seeking

other destinations. The break-up of Ethiopia, which led to the creation of Eritrea, also has its origins in Cold War superpower rivalry. Indeed, even the Soviet exploitation of Ethiopia after the 1974 coup against the Emperor was very much a part of the Cold War;

- The ongoing conflict within South Sudan and the prospect of unresolved issues between South Sudan and Sudan remains of great concern, despite the signing of the August 17, 2015, Addis Ababa peace accord between rival South Sudanese factions. That difficult process was a great achievement for IGAD, but it is clear that the underlying issues of that conflict have not been resolved;
- Internal conflict in Uganda and neighboring states still exists as a result of the activities of the Lord Resistance Army, and so on.

There are, in addition, numerous other simmering issues of concern, not least the sharing of the waters of the Blue and White Nile rivers, the unresolved disputes between Eritrea and Ethiopia, the unresolved issue of the sovereignty of Somaliland, and more. There is also the impact of the pressure exerted by external players, from the US and France to the People's Republic of China, in Djibouti; the Turks in Somalia; the Qataris in Eritrea and Sudan; and so on and so forth.

We have seen the secondary consequences of these Cold War and post-Cold War conflicts — whether in Somalia, Sudan and South Sudan, or Libya — which have exported terrorism, insurgency, and crime into other parts of Africa.

The current conflict against *Boko Haram* in north-eastern Nigeria, for example, still draws sustenance from the safe haven of radical groups in Somalia and Sudan, quite apart from the involvement of the ongoing instability in Libya which has provided a conduit for movement of radical fighters and weapons into West Africa. We cannot forget that this is directly the result of ill-conceived intervention by the major powers in the Libyan political crisis which triggered off when the Constitutionalists in Cyrenaica began a protest against Libyan leader Mu'ammar al-Qadhafi in 2011.

So these current challenges have much of their origin in conflicts, rivalries, and ideologies of Cold War competition between the West and the Soviet *bloc*, or were caused by the post-Cold War power vacuum and the revival of competition by major powers. This is now changing much of the strategic reality in Africa and the Middle East.

So, what the region is facing today, and which is having an impact on much of the Continent, is not the legacy of colonialism but the legacy of the Cold War and competition between the major powers. Today, that "major power competition" involves state and non-state actors, and new forms of warfare.

What all this means is that the nature and the framework, or context, of the security challenge has changed, and is likely to evolve still further in the coming years.

If the threat and the context are changing in their very nature, it is logical, too, that we see a shift in the way in which we in the region shape our political, diplomatic, and military institutions to deal with problems.

The international community — both the great and interested powers outside Africa, as well as the United Nations — need better understanding, intelligence, and security mechanisms to help us restore stability and prosperity in the region. The architecture of global power is transforming rapidly. Engagement of great powers in Africa and the Middle East involves different attitudes and different players; while the framework and possibilities that exist within the Horn, Red Sea, and Middle East are changing. It is clear, therefore, that it is time for a full re-evaluation of the situation as well as our mechanisms for managing sovereignty and our interaction.

We, in the region, must be aware that what we are calling "the international community" — the industrial powers outside Africa and the Middle East — do not devote a major proportion of their diplomatic or intelligence skills to the region under discussion: the Red Sea/Horn of Africa region. They perceive that they have other priorities.

As a result, African and Middle Eastern states have to ensure that they have appropriate, contextual, and sophisticated intelligence and analysis to ensure that not only can the region master its own problems, it can also promote better understanding of the issues by our international partners and relationships that are appropriate and effective.

The questions which must be posed to the African Union, and for the regional security bodies such as IGAD, as well as to the Arab League (which has interests in North Africa and the Horn of Africa), include the following:

1. Does "peacekeeping", as it is presently construed by the United Nations concept, really provide long-term resolution of the problems which caused the conflicts? In other words, have we been embarking, for the past seven decades on a process which merely suppresses immediate fighting without addressing the need to find just and workable solutions which would endure for the longer-term?
2. Are African political, diplomatic, economic, and security mechanisms for conflict prevention adequate, given the changing global balance and transforming social and conflict patterns?
3. Are our intelligence collection and sharing mechanisms adequate at national and at regional levels? And, as a direct corollary of that, are our threat assessment and

warning mechanisms adequate?

4. Are our regional conflict resolution approaches adequate? Quite apart from our diplomatic channels, when we get to the physical phase, are we using our multinational coalition forces effectively, both within the region and with regard to the United Nations? We are currently working with regional peacekeeping/conflict suppression mechanisms such as the East Africa Standby Force (or the other four such forces on the Continent), as well as Continent-wide AU bodies, and UN forces, and then a range of other coalitions. Are these military mechanisms really viable? Are they, too often, too blunt? Too late? Do they merely — as I said — attempt to suppress immediate violence and yet, in so doing, prevent identifying and remedying the root causes of the conflicts?
5. Are the African Union's and the Arab League's political-diplomatic capabilities adequate to engage the international community appropriately and with sufficient intelligence, bearing in mind that this is a region — the Suez/Red Sea/Horn — which contains essential global assets? Certainly, the great powers as they are now emerging have no intention of leaving the security of this region solely in the hands of the local states. And if some of the "great powers" have reduced their presence, it is clear that others, including the People's Republic of China, and perhaps India and, once again, Russia, are intent on increasing their engagement.

The answer to the question of the adequacy of our existing capabilities is clearly in the negative. We would not be facing the security challenges we now face if our capabilities were sufficient.

Other than, for example, adding satellite and unmanned aerial vehicle technologies, or cyber capabilities, what has changed in our approach to security and conflict prevention in the region in the past half century? The answer is very little.

What, then, must be done? And which of our existing skills and capabilities need to be reinforced to achieve strategic objectives in the region?

To begin with, perhaps we need to raise the level of responsibility of the regional bodies, such as the African Union, IGAD, and the Arab League. These bodies have, since World War II, deferred to the UN as the ultimate arbiter, but perhaps the UN should now see that real responsibility should be devolved to a greater extent on the regional bodies.

But if we are to follow that path, then clearly the level of thinking, and expertise, needs to be significantly improved at the local and regional levels. Already we see in Somalia that the African Union is taking much of the responsibility.

The South African Institute for Security Studies, in a report on October 5, 2015[10], noted: "At the moment, the European Union almost singlehandedly finances the very costly AMISOM. However, earlier this year, it announced that others will have to step up to the plate and that it is going to reduce its contributions. 'Where are the other countries who are also concerned? The Gulf States, Saudi Arabia?' asked a European diplomat in Addis Ababa recently. 'China is not contributing much either. They need to enlarge the circle of those paying for the AU to do the job,' said the diplomat. The ideal for the AU would be a steady stream of funding from assessed contributions of all member states of the UN."

The report went on to note: "[The People's Republic of] China's ambassador last month handed over a check of $1.2-million to the AU Commissioner for Peace and Secu-

10 See Liesl Louw-Vaudran: "Africa does its bit — and more — for peacekeeping", in *ISS Today*, October 5, 2015. https://www.issafrica.org/iss-today/africa-does-its-bit-and-more-for-peacekeeping.

rity Smail Chergui at a ceremony in Addis Ababa. The contribution pales in comparison to the nearly $20-million per month that AMISOM costs, but it is a start."

For Africa to expect the World to continue to pay for peace on the continent forever is unrealistic.

Again, this gets back to some fundamental questions: Have the resources provided achieved desired results? Are we doing our peacekeeping correctly if we are to achieve long-term normalcy, stability, and growth? What are our objectives? Have we even defined them?

Are we struggling to preserve national boundaries? If so, should we not acknowledge that some of our original goals in preserving existing nation-state borders have already been reconfigured in the region in the recent past? For example, Sudan is now in two parts.

The union of the former British and Italian Somalilands into Somalia in 1961 has ended, with the former Italian Somaliland being now regarded as "Somalia", while former British Somaliland is now, once again, the Republic of Somaliland. Ethiopia broke into two components, Eritrea and Ethiopia, following the end of the Cold War. And, following the union of the Yemen Arab Republic (North Yemen) and the People's Democratic Republic of Yemen (South Yemen) in 1990, we now see Yemen once again facing the prospect of re-dividing into "north" and "south", or perhaps into even smaller constituents, reflecting the diversity which preceded the British occupation of southern Yemen and the Hadhramaut in the mid-20th Century.

The 21st Century is likely to be at least as dynamic as the 20th Century in the creation of new states and the transformation or end of some older states. We have, in the African Union, seen a gradual modification of the old Organization for African Unity stricture that the boundaries of African states, as defined by the colonial era, should remain sacrosanct. There is an attempt to accommodate changing realities, and the AU's Border Program has been laying out legally acceptable guidelines for that[11].

The truth is that throughout history, including African and Middle Eastern history, boundaries and the existence of states have been evolving. We should probably expect that this will continue to be the case, and our responsibility is to ensure that the change occurs in a manner which is manageable, beneficial to all parties, and produces self-sustaining societies.

Perhaps I am raising more questions than I am answering, but these are questions which are rarely and comprehensively faced.

Part of what we lack has been the ability to achieve an ongoing program of goal-setting, either at national or regional levels. Surely that must be a starting point.

Another aspect of our process has been that we have based our processes on a continuation of the orthodoxy of the 1945 world order, as defined by the victors of World War II in creating the United Nations. This has been inordinately successful in stabilizing the global framework, but change is now occurring at a rapid pace, and Africa and the Middle East — and other regions — must now also see that they have a responsibility to ensure that their own processes are appropriate for the changing times.

Within our present framework, there has been a universally-accepted approach, since 1945, to international peacekeeping philosophies. Perhaps it is time to transform this more into a more holistic "conflict resolution" process, which addresses underlying causes to a greater degree. Our historical reliance merely on stopping fighting in the short-term has failed to resolve problems, and this failure has condemned whole societies to a limbo of paralysis,

11 See: *Creation and Operation of Boundary Commissions in Africa: The User's Guide. African Union Border Programme (AUBP)*, Second Edition, August 2014.

where they cannot go forward. This, of course, exacerbates the impetus for people to flee their homelands in migratory waves, and to leave those homelands condemned in frozen conflicts, never, it seems, to achieve resolution, happiness, or prosperity.

What we have seen in the past year or so in the Suez/Red Sea/Horn of Africa situation is, to be sure, the continuation or revival of some conflicts. But what we have also seen is a new climate of cooperation which has occurred outside the framework of the old global balances of power.

We have seen Egypt and Ethiopia begin to cooperate, and this could truly benefit the region and the stability of the Red Sea international sea lanes. We have seen the prospect that Ethiopian-Sudanese cooperation on the generation of hydro-electric power, with the prospect that this could extend further. We have seen the growth in Ethiopia-Djibouti commercial and social linkages. We have seen the real transformation of Egyptian-Israeli commercial ties, with further linkages into the evolution of an Eastern Mediterranean energy basin. We have seen the unilateral and substantial Egyptian investment — not only by its government but by its people — in greatly expanding the capacity of the Suez Canal, to the benefit of the entire region and the global trading community.

These things occurred because of the bilateral, or at time multi-lateral, initiatives of the various leaders and governments involved, rather than through international organizations.

So the trends are, in many respects, positive, and we can say this without diminishing our concerns over the ongoing challenges caused by conflicts in Yemen, Somalia, and so on, and the overflow of conflicts just outside the boundaries of this discussion.

What, then, should our next steps be for this vital region, on which so much of the fortunes of Africa and the world depend? With great respect to our existing regional bodies, most of which do not specifically address the unique Suez/Red Sea/Horn region; and which are often, of necessity, reactive as means of addressing issues as they arise; and in some instances do not include all the players, let me suggest the following:

First, the region — that is, the Suez/Red Sea/Horn of Africa strategic zone — does not appear to be sharing intelligence and thinking on anywhere near the appropriate level of quantity, quality, or timeliness.

Secondly, the region has been reluctant to develop an overarching framework in which mutually-beneficial goals could be created and developed on an ongoing basis.

And Thirdly, progress has been hampered by existing diplomatic formalities which often serve to preclude the open exchange of ideas, intelligence, and analysis between all engaged parties (some of which are not recognized formally by others).

To start the process of resolving this, I recommend that we allow the creation of an informal clearing house for the exchanges of the kind we presently lack, and which I have just referenced. This "clearing house" could be the discreet channel by which the regional players can raise and discuss ideas, not only with each other, but with the international community.

This should be a low-cost, "good offices" mission, managed by an engaged, but non-partisan professional body, administratively functioning outside the region so as to avoid bias. In this regard, I suggest that this should be the International Strategic Studies Association, based in the Washington, DC, area, and I would also add that my own organization, the Gusau Institute, which is African, but not of the Suez/Red Sea/Horn region, is also prepared to support and be actively engaged.

If we look at the region's littoral and dependent states — Egypt, Sudan, Eritrea, Israel, Jordan, Saudi Arabia, Yemen, Djibouti, Somaliland, Somalia, Ethiopia, and South Sudan — then funding by each of a few hundred thousand dollars a year would create a sufficient resource designed to benefit them all, as well as benefit all of Africa, and the global trading community.

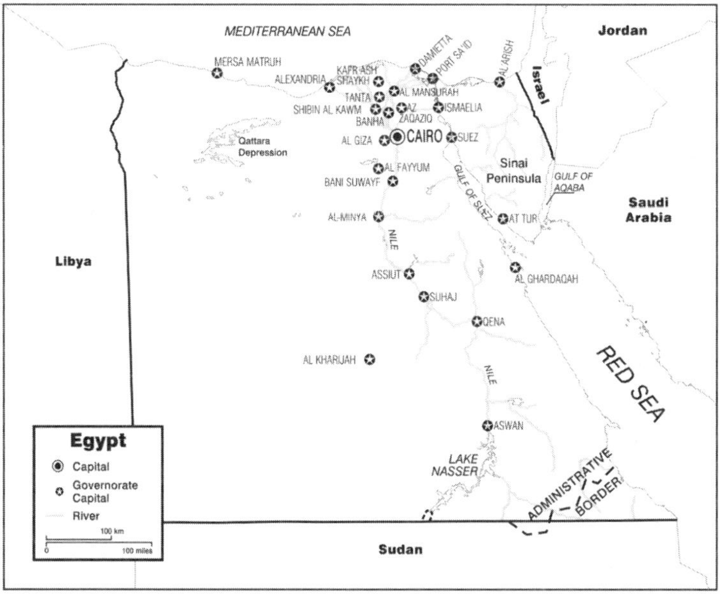

Two Egyptian Air Force AMD *Rafale* fighters acquired in 2015 as Egypt began to diversify its defense procurement sources.

V

Egypt's Return to Strategic and Economic Centrality

Gregory R. Copley

EGYPT IS NOW WELL INTO THE PROCESS of rebuilding its historically-significant capabilities and its prestige. It is also rebuilding the strategic balance it has with the region, and it is rebuilding its internal balance. Much of this has to do with the fact that the Cairo-Washington relationship abruptly cooled in 2014 (even though in 2015 it resumed some of its warmth).

The four decades or so of US-Egyptian closeness ended without fanfare, even without the recognition in Washington that the event had occurred. Whatever happens now, that era of closeness between the two states is unlikely to be repeated.

Cairo regards its relationship with the US as important, but has now also committed itself to a broader spectrum of international relationships to stimulate its security and its economy.

This has already begun to take effect.

Domestically, on October 18 and 19, 2015, Egypt began the first round of its legislative elections, and concluded them on November 22-23, 2015, leading to the re-establishment of a new Parliament co-equal with the Presidency, a Presidency with fixed terms of office. This was just one of the visible signs of change, but meaningful change has been underway for some two years: energy subsidies have been reformed, the tax system improved, the budget deficit reduced, and two stimulus packages of $8-billion have been injected into the economy.

The August 2015 opening of the expanded Suez Canal, a massive achievement on its own, was only a first step toward transforming the new global logistics hub and industrial processing area to serve European, Asian, and Persian Gulf markets, potentially creating a million new jobs. The Suez Canal Area Development Project (the SCZone) includes the building and re-development of four major ports — East Port Said, Ain Sokhna Port, 10th Ramadan Dry Port, and Ismailia Dry Port — which will all be powered through new wind farms, gas turbines, and waste-to-energy incineration plants.

The Cold War was over when, in 1995, my book, the *Defense & Foreign Affairs Handbook on Egypt*, was published, noting in its introduction:

> Egypt has always given the world insights into the lessons of history, should we care to learn from them, and the tools to move forward to secure a more positive future. It is not surprising as the world moves from the stable, superpower-dominated Cold War era to a more fluid, multipolar strategic environment, that Egypt should be the key to much of what will happen in terms of global security in the coming decades. The country is geopolitically placed at the nexus of trade routes, as well as the hub of several cultures and religions.[12]

It has taken 20 years for that observation to become fully-formed and starkly real, and yet we still see that there

12 Copley, Gregory R.: *The Defense & Foreign Affairs Handbook on Egypt.* London, 1995: International Media Corporation.

is insufficient recognition of this geopolitical reality. Indeed, we are only now starting to see, in the post-post-Cold War world, a revival of geopolitics as a significant prism through which to view the world.

Egypt, on October 15, 2015, was elected, against heavy competition, to the vacant two-year seat representing Africa on the United Nations Security Council. It was a measure of how far Egypt's fortunes have revived; it received 179 out of 193 votes, on this, the fifth time Egypt has been elected to a non-permanent seat on the Security Council. Its two-year tenure began on January 1, 2016.

Every component of the Suez / Red Sea / Horn of Africa matrix of nation-states is a significant element in the overall framework, but Egypt has throughout history been the anchor of the region. This is not merely because it owns the Suez Canal and the attendant geography which places it at the center of the Eurasian-Africa connection, but because of its historical civilization and traditional leadership rôle (which today contributes national identity and confidence, and tourism), as well as the depth of its educational, manufacturing, and agricultural fabric. To that mix has now been added a highly-significant natural gas sector.

But what is critical in re-evaluating Egypt's rôle is to understand that Egypt's context has changed. We saw a significant reduction in the relative strategic maneuverability of most of the region's nation-states during the past couple of centuries, because of the overriding influence of major powers, from Britain, France, and Turkey, to the US and the USSR. But what is critical today in looking at the relative weights of the players in the Mediterranean-to-Indian Ocean theater are the changes wrought by new economic factors and now, particularly, the reassertion of geography as a driver. Some of this also draws on the foundation of the normalization of relations with Israel, which Pres. Anwar as-Sadat achieved in another interregnum between superpower engagement.

In the immediate perspective, Egypt's economic growth rate has returned for fiscal 2015-16 to more than four percent per annum, after a decline during the period of the Government of Pres. Mohammed Morsi. Egypt is now poised, with the economic reforms underway, to grow at an even faster rate. The successful outcome of the March 2015 Egypt Economic Development Conference secured more than $60-billion worth of investments, loan agreements, and grants, and reinforced the Government's commitment to continue the structural reforms and promote inclusive growth and restore investor confidence.

Egypt's GDP in 1993 was some $65.5-billion, which, adjusted for inflation to 2014 dollars, would be some $111-billion; and its population around 57-million. By 2014, the GDP was $286.54-billion, and its population was 88.4-million. In other words, its population has multiplied by 155 percent, and its GDP by 258 percent in that same period, a clear increase in *per capita* wellbeing.

Egypt's economic revival has been substantial and linked to its reforms. In the first year of Pres. Sisi's Government, tax revenues rose by $39.5-billion in the first six months of FY2014/15, to $231.3-billion, a 21 percent increase over the same period a year earlier. And within that first year of the new Government, the Big Three rating agencies upgraded Egypt and positively revised their outlooks.

The oil-dominated regional economy of the post-Cold War period had, for a period, relegated Egypt to a comparatively lesser status. It had, to a great extent, become a precarious economy, and, in security terms, a client state of the United States. The transformation of the global energy marketplace has now diminished the political power of the major oil producer states, therefore, in relative terms, raising Egypt's standing. The series of regional missteps of the 21st Century by the major Western states, particularly the US, led to an increasing vacuum of external power in the region, which was concurrent with strong US, Turkish, and Qatari support for the installation of a Muslim Brotherhood government in Egypt.

There is no need to dwell here on the process which led to the installation in mid-2012 and subsequent public rejection in mid-2013 of the Muslim Brotherhood's Freedom and Justice Party (FJP) (*Hizb al-Hurriya wa al-'Adala*) Government of Mohammed Morsi, other than that the early refusal of the US White House to accept the Egyptian rejection of the Muslim Brotherhood caused a watershed in US-Egyptian relations. Moreover, the continued US, Turkish, and Qatari insistence of support for the now-exiled Muslim Brotherhood — which remains committed to an armed struggle against the Egyptian State — has made Cairo cautious about Washington, and hardened against Turkey. There is no doubt that the Turkish Government under Pres. Reçep Tayyip Erdoğan remains committed to working against Egypt. This is a significant, but rarely discussed, factor in Egypt's security framework, and there was hope that this would diminish if Turkey's November 1, 2015, elections had removed Pres. Erdoğan's AKP from control of the Parliament. This did not happen, and, in fact, the AKP regained dominance in the Grand National Assembly, and the President resumed his unconstitutional dominance of the Government, and strengthened his commitment to support regional Muslim Brotherhood bodies, including those of Egypt.

As a result of the process to this point, however, Egypt had begun, by 2014, to strategically distance itself somewhat from the United States and Washington. Many in Washington did not initially seem to even understand that the seismic shift had been made. Egypt had begun, in 2014, reviving economically and strategically, and promised to transform the Suez-Red Sea strategic maritime environment. Meanwhile, the US White House still supported the opponents of Egypt — particularly Turkey, Qatar, and the Muslim Brotherhood — while now attempting to rebuild US-Egyptian relations, particularly with the August 2, 2015, US-Egypt Strategic Dialog.

Even by late 2015, the new strategic reality in the Middle East still did not appear to be grasped fully in Washington: Egypt had broken its strategic dependence on, but not its friendship with the United States. And, once again, few in Washington noticed. Once again? Let us begin with the first episode.

It was on April 9, 1972, that I debriefed a source who had returned to the US from Cairo with startling news from his meetings with senior Egyptian Defense Force commanders: The Egyptian Government of Anwar as-Sadat would terminate its strategic relationship with the Union of Soviet Socialist Republics (USSR) within six months, and all Soviet advisors would be asked to leave Egypt. And yet Egypt was, at that time, 100 percent dependent on Soviet military equipment, advice, and aid.

The source was impeccable; his access was unquestionably top-level (and his relationships in Cairo were on a personal level; he was not a US Government official). The information was used in the second edition of our weekly *Defense Newsletter* briefing, published on April 10, 1972. It was distributed to the yawning disinterest of some and the angry incredulity of others in Washington, DC: This could simply not be the case; Egypt was locked in to the Soviets and had no recourse or strategic maneuvering room. Egypt, for Washington, was irretrievably in the enemy camp.

Less than four months later, on July 18, 1972, Pres. Anwar

as-Sadat did indeed evict the Soviets. The US was, in any event, preoccupied and still failed to pick up the threads. US Secretary of State Henry Kissinger was deeply engaged in negotiations with North Vietnamese officials to end the Vietnam War.

Just over a year later, on October 6, 1973, Egypt launched its military offensive against Israel — with its existing Soviet equipment, but without Soviet support and without any supplies of spare parts or Soviet servicing for more than a year — to recover control of the Suez Canal and the Sinai. The war transformed Egypt. The US — with the Camp David Accords signed on September 17, 1978, following Anwar as-Sadat's peace offensive with Israel and his visit there in 1977 — began to pay attention to the reality that Egypt was not in "the enemy camp".

What Pres. Sadat had successfully done was undertake a two-stage transformation of Egypt's strategic position in such complete secrecy that, despite the massive, national-level nature of the preparations for each stage, had gone unnoticed by the Soviets, the US and NATO, and by Israel. When he inherited the leadership of Egypt, with the death of Gamal Abdel Nasser in 1970, the nation was still in the grip of depression as a result of the humiliating Six Day War with Israel (June 5-10, 1967). Sadat inherited a military force in which just six percent of its officers had a university education.

By 1973, when the next war began, more than 60 percent of Egyptian officers had received, or begun, university education. The Egyptian military had gone from a system which had totally eschewed initiative at all levels to one which began to think.[13]

Sadat transformed the global strategic framework by

13 For a more comprehensive coverage of the transformation see, Copley, Gregory R.: *The Defense & Foreign Affairs Handbook on Egypt*; London, 1995: International Media Corporation.

moving dominance over the Red Sea/Suez from Soviet influence and ensuring that the Suez could re-open; he transformed Egypt's economic and military framework; and he achieved peace with Israel. And no major foreign power knew what he was planning until he executed his strategies.

Egypt's current President, Abdul Fatah al-Sisi, has demonstrated that he has Sadat's strategic grasp and courage. He moved on July 3, 2013, to terminate what had been a blatant distortion of a democratic process which — with the strong insistence of the US Barack Obama White House — had put in place a Muslim Brotherhood Government under Pres. Mohammed Morsi. In subsequent elections, on May 26-27, 2014, former Gen. al-Sisi was overwhelmingly elected to the Presidency.

Pres. al-Sisi has transformed Egypt, while Washington, essentially, ignored him. Apart from significant economic initiatives to revive the country after a devastating year of collapse under Morsi, the Government of Pres. al-Sisi has begun to rebuild Egypt's strategic posture and future, essentially operating on the premise that Washington would, or might, no longer be a supportive ally.

To begin with, a new strategic framework has been emerging in the region, with a *bloc* consisting of Egypt, Israel, Jordan, Saudi Arabia, Kuwait, and the United Arab Emirates. Egypt and Israel have the capacity to call on Russia; Israel has increasingly close relations with Greece and Cyprus, and Egypt, too, has been working with Israel, Cyprus, and Greece on the exploitation of hydrocarbon resources in the Eastern Mediterranean.

In October 2014, the Government signed the initial contracts on the $8.4-billion project which was designed to, within less than one year, deepen and widen the Suez Canal. This included the construction of a 50km parallel section of the canal to allow ships to transit simultaneously in both directions, and widening and deepening a number of existing sections to a depth of 24m. Despite economic and security concerns, the Government took the risk to deliver the finished, wider canal for use by August 2015. It was completed on schedule, and stands to significantly increase the income potential of the Canal, Egypt's most significant earner of foreign exchange, by expanding the through shipping from 49 to 97 ships a day.

Once it became clear that the Obama White House was using its military supply relationship with Egypt to attempt to enforce a relaxation of the Government's ban on the Muslim Brotherhood, the Sisi Government began to discuss the prospect of buying non-US military systems.

US officials, when told of this development in 2014, laughed off the suggestion that Egypt could so easily replace — or operate without support — its extensive US inventory, including its 250 or so Lockheed Martin F-16 combat aircraft variants, its almost 1,500 M-1A1 *Abrams* and M-60A1 main battle tanks, and other systems.

They forgot how smoothly Sadat had replaced the Soviets, and that he had been able to continue military operations without immediately finding a new supplier state. The significance of the Egyptian military industrial community, particularly centered around the Arab Organization for Industrialization (AOI), was critical for that process.

Discussions with Russia for major defense systems had just begun in late 2015 to yield results, but the Sisi Government had by that time begun negotiations with France for major combat systems. Egypt did commit in early 2015 to buying its first nuclear reactor from Russia. Much of these negotiations were delayed by the complexities of financing arrangements and Egypt's struggle to rebuild its economy. Overall, Egypt's economy benefitted from major new foreign investments, loan, and aid. The governments of Saudi Arabia, UAE, and Kuwait had immediately pledged nearly $20-billion in financial aid when Gen. al-Sisi won the presidency, and $12-billion was reportedly made available almost immediately.

US aid to Egypt under the Camp David accords amounted to only $1.3-billion a year.

In 2014-16, the Egyptian Armed Forces significantly diversified their procurement pattern with the following acquisitions:

- A total of six new, very advanced *Gowind 2500*-type corvettes from France;
- Two new NATO FREMM-type 6,000 tonne displacement frigates from France;
- Twenty-four fourth-generation French AMD *Rafale* B and C fighters, including some *Black Shaheen* air launched cruise missiles;
- Two new 21,300 ton displacement *Mistral*-class helicopter assault ships from France; ships which had been built for Russia, but which could not be delivered because of embargoes;
- Fifty Russian Kamov Ka-52 *Alligator* attack helicopters, including the naval variant, the Ka-52K *Katran*, for use aboard the *Mistral* assault ships;
- Negotiations were underway to acquire Russian S-300 advanced surface-to-air missile systems;
- Negotiations were underway to acquire a substantial number of MiG-29 or MiG-29-derivative fighters (MiG-35s) from Russia, to replace EAF MiG-21s[14];
- Discussions were underway for possible procurement of Pakistan-built JF-17 *Thunder* Block III fighters (with aerial refueling capacity) to replace EAF/ Chengdu J-7 and AMD *Mirage V* single-engine fighters.

14 Some sources suggested that the initial EAF order could be for 48 MiG-29SMTs and 24 MiG-35s, or 46 MiG-35s and 26 MiG-29SMTs; other reports have indicated that an order could be placed for 62 MiG-35s. The order was expected (as of January 2016) to be in excess of $2.2-billion.

Symbol of Continuity: The Egyptian State Yacht *Mahrousa* enters the Suez Canal for the historic re-opening in August 2015. It was the third time the ship had led the opening of the Canal since 1869.

The Sisi Government wished to utilize the unveiling of the massive increase in capacity of the Suez Canal — funded by Egyptian public subscription — as a major step forward for Egypt, similar to the opening of the original Suez Canal in November 1869. Isma'il Pasha, the *Khedive* (ruler) of Egypt and Sudan, utilized the Canal opening to portray Egypt as a transformed strategic entity. Giuseppe Verde's opera, *Aïda*, was written to celebrate the Canal's opening, and its opening performance was at the Khedivial Opera House in Cairo on December 24, 1871. The greatest yacht in the world — the *Mahrousa* — was built in London so that the *Khedive* could transit the Canal for the opening ceremonies, which were, rightly, of paramount interest to the global maritime community. Significantly, *Mahrousa* carried Pres. Sisi through the new section of the Suez Canal when it was opened in August 2015.

[It is not a coincidence that Saudi Arabia also remained cool to the US and it, too, in October 2015, said that it would open talks to buy French reconnaissance and communications satellites, patrol boats, and the setting up of a naval research center and a working group to study future naval projects. That commitment from the Kingdom was reportedly to be worth some $2-billion, and be concluded by the end of 2015. It would also include the acquisition, worth some $600-million, of 30 French-built 30 meter patrol boats. Certainly, Saudi Arabia is continuing some US defense buys; it had, by November 2015, negotiated the $11-billion acquisition of four US Lockheed Martin *Freedom*-class Littoral Combat Ships (LCS) (which the Saudis designate as MMSC: Multi-Mission Surface Combatant), for the Royal Saudi Navy's Persian Gulf fleet. However, it was also negotiating to buy two French frigates for the Red Sea fleet. Saudi Arabia also finalized details of a deal worth $1.9-billion to buy 10 Sikorsky MH-60R *Seahawk* helicopters, for anti-submarine warfare and other maritime duties.]

Egypt will clearly go much further down the path of reasserting its strategic independence, and now has the education base, the commercial sector, and the energy production which was not available to Pres. Sadat when it built its sophisticated defense industrial base, with the help of other Middle Eastern states (particularly Saudi Arabia).

Egypt's Arab Organization for Industrialization (AOI), its main defense industrial sector, which has designed and built aircraft on its own, has also teamed in the past with Dassault (AMD), of France; CATIC, of the People's Republic of China; and Embraer, of Brazil, to build aircraft. So it would not be surprising to see Embraer's new KC-390 tanker-transport aircraft come under consideration to replace the Egyptian Air Force's Lockheed Martin C-130H *Hercules* in the foreseeable future. It is possible that consideration could also be given to the Airbus A-400M tactical transport aircraft.

AOI facilities are almost certainly taking up the slack in supporting Egypt's now active air combat operations against Islamic State operatives based in neighboring Libya and in the Sinai.[15]

Egyptian Air Force (EAF) F-16s struck Islamic State bases and fighters in Derna, Libya on February 16, 2015, in response to the beheading by ISIS militants of 21 Egyptian Coptic Christian workers. Significantly, the US has refused to support the Egyptian actions against ISIS, which reinforces Cairo's belief that it must be able to safeguard its interests unilaterally, if necessary.

The brief 2013-2014 White House rift with the Egyptian Government flew in the face of historical US interest in maintaining freedom of navigation on the global maritime commons. The rift threatened the US ability to have influence in the Suez Canal/Red Sea SLOC. And the White House action seemed predicated on ideological and religious grounds, rather than on US strategic interests. The resumption of the Egypt-US Strategic Dialogue in August 2015 brought US Secretary of State John Kerry back to Cairo to help bring the relationship back to one of warmth.

But what is certain now is that the US has ended its almost four decades of strategic dominance in Egypt, even though the friendship and a strong security relationship will continue. This, with attendant proportionate reduction in US influence in the Mediterranean-Suez-Red Sea-Indian Ocean SLOCs, may be the most significant erosion of the US global posture in recent years.

The Egyptian Navy was, by 2016, looking to start expanding its submarine force, a long-overdue requirement to increase its fleet of two German HDW Type 209 submarines, which had been expected to be supplemented by two more of the same type. The PRC, which held joint naval maneuvers with the Egyptian Navy in September 2015, has

15 See: "Libya, Now Effectively Divided, May Be Focus of Major Rift Between US and Regional Allies", in *Defense & Foreign Affairs Special Analysis*, August 27, 2014. Also "Libya, Now Effectively Divided, Seeks a Way Forward", in *Defense & Foreign Affairs* ***Strategic Policy***, 7-2014.

offered the S-20 export variant of the Type 41 *Yuan*-class diesel-electric submarine, possibly equipped with Air Independent Propulsion (AIP). AIP may be a cost-prohibitive step too far for the Egyptian Navy at present.

These are the same type of submarines which Beijing has under contract for the Pakistan Navy, which has ordered eight of the class. There is a pattern here: the PRC views Pakistan as a stable partner in protecting its Suez-Red Sea-Indian Ocean maritime trade. It is axiomatic that the PRC has a clear interest in maintaining and supporting stability and security of its sea trade all the way to the Mediterranean, and a sale of S-20s on favorable terms to Egypt would be a significant step in this. The PRC did, decades earlier, provide the four *Romeo*- or *Ming*-class SSKs to the Egyptian Navy; the history has already begun.

What continues to be demonstrated, then, is that Cairo is making its strategic decisions well outside the basis of its recent relationship with the US. This relationship had dominated its military policy planning from the time of the Camp David Accords in 1978 until the Obama White House essentially sabotaged US-Egyptian relations in 2013 by supporting the Muslim Brotherhood Government of Pres. Morsi. What is significant is that the Egyptian Government clearly seeks to retain a positive relationship with the US, so the latest break is not similar to Pres. Anwar as-Sadat's break with the USSR in 1972. But even then, Pres. Sadat did not entirely break relations with Moscow in favor of the US or the West.

Meanwhile, in early December 2015, very senior governmental sources in Moscow and Cairo confirmed that Egypt was in the very late stages of negotiating a contract with Russia for the supply of a "significant quantity" of MiG-29 fighter aircraft, possibly even the MiG-35 advanced variant of the MiG-29. It had been reported in late September that the Egyptian Air Force was considering a buy of 46 MiG-35s from RAC MiG for approximately $2.2-billion. Some sources have put the initial Egyptian request at 62 MiG-35s.

Russian sources referred to the proposed buy as "MiG-29s". The deal could be signed at any moment, the sources indicated. The website AIRheadsfly.com indicated in August 2015 that it would not "rule out a combination deal: 48 MiG-29SMTs and 24 MiG-35s — or 46 MiG-35s and 26 MiG-29SMTs".

There was no suggestion that the new aircraft buy was intended to replace the advanced models of the existing backbone force of Lockheed Martin F-16 variants, although clearly a significant number of MiG-35s would allow for the replacement of earlier F-16A/B Block 15 fighters, and its MiG-21 and Chengdu F-7 fighters.[16]

The acquisition of MiG-35s (or even a slightly less advanced MiG-29 variant) would provide a significant boost to the top-end capability of the EAF, which already has acquired the first of its 24 AMD *Rafale* B/C fighters, and would significantly broaden its security from what had been a very strong reliance on the F-16 (and therefore on the US) for credible air power stability. In the lower end of the fighter spectrum, the EAF has a significant component of AMD *Mirage 2000* variants and Chinese Chengdu F-7 variants. The EAF has been extremely successful in maintaining the operational readiness, too, of even older fighter types, such as its McDonnell Douglas F-4E *Phantom IIs* and

16 The Egyptian Air Force (EAF) F-16 inventory is roughly as follows: A total of 260 delivered, or in the process of delivery, of the following variants: 33 Lockheed Martin F-16A Block 15 tactical fighters; 5 Lockheed Martin F-16B Block 15 operational trainers less than 34 Lockheed Martin F-16C Block 32 tactical fighters; less than 6 Lockheed Martin F-16D Block 32 operational trainers; 78 Lockheed Martin F-16C Block 40 tactical fighters; 27 Lockheed Martin F-16D Block 40 operational trainers (total of 220 F-16s delivered as at April 2002); 24 Lockheed Martin F-16C/D Block 50/52 fighters (ordered, Oct. 2009; 16 del. by Jul. 30, 2015; remainder by year-end 2015). Source: *Defense & Foreign Affairs Handbook Online* and the Global Information System (GIS).

some of its MiG-21s, to ensure it has a back-up capability available.

What is significant in the proposed MiG transaction, which had already been under consideration and negotiation for more than a year, was that there had been little evidence that the US Government had permitted or encouraged Lockheed Martin to pursue a continuation of F-16 sales or consider the sale of F-35s (which have been specified by, for example, the neighboring Israeli Air Force). The US Obama Administration has, to a significant degree, then, allowed itself to retire from attempts to woo Egypt back into the US strategic fold, even as Russian, French, and PRC moves have made significant inroads into the Egyptian defense spectrum.

The consideration of a buy of Pakistan Aeronautical Complex JF-17 *Thunder* Block III fighters would be intended to replace some or all of of the EAF's remaining 50+ AMD *Mirage 5* fighter variants, and the remainder of its fleet of 45 Chengdu F-7B tactical fighters.

British Parliamentary and defense industry sources by late 2015 had also expressed concern that the UK Government had failed to be aware of the change in Egypt's strategic alliance posture over the past year or more and had therefore missed the opportunity of selling British defense systems, including the Eurofighter *Typhoon II* fighter, to Egypt. The UK has had considerable input into the Egyptian defense industrial capability in the past, but gradually allowed itself to be eclipsed by Egypt's growing alliance with the US during the Administration of Egyptian Pres. Hosni Mubarak (1981-2011).

Egypt's recent polite break with the US — triggered by the Obama Administration's failure to read the political methodology of Egypt which had been clear since the fall of the Monarchy to the Free Officers' Movement in 1952 —

was inevitable, given the US decision to withdraw from its earlier military-oriented dominant position in the Middle East in favor of what it had hoped would be a soft-power approach. That was in large part the reflection of the exhaustion by the US with the Iraq and Afghanistan wars. But Cairo was clearly ready for change, and, as well, the parting gesture by Washington to support an assumption of power by the Muslim Brotherhood was not something the US was able to enforce. So when the Egyptians rejected Morsi, and when the US was unable to prevail in its insistence on Morsi's retention, the emerging Egyptian Government was on a path for a new life outside the constraints of a superpower guardian.

Where does Egypt go from here?

It has moved in an incredibly short period of time through the phase of being a component of the US network in the region; through the economic and social disruption of the Morsi period; to a position of growing prestige, capability, and promise. It is, *de facto*, part of a network of economies which would be expected to work increasingly closely, and include Israel and Jordan, Saudi Arabia, and the UAE. Egypt has a strong stake in ensuring the continued stability of Saudi Arabia, and in restoring calm to Yemen.

Cairo is also continuing to build its focus on Africa, and the fact that it has won the Africa seat in the UN Security Council reinforces this. The African Union Summit in January 2015 endorsed the Egyptian initiative to develop mediation mechanisms within the AU to better address conflict prevention and resolution efforts on the Continent. Egypt is also using its soft power as a moderate Muslim African country to fight the rising threats of extremism. Egypt's top Islamic institutions, Al-Azhar and Dar-Al-Eftaa play a rôle, dispatching delegations of clerics to the Sahel and East Africa, and developing a counter-narrative to the extremists in local African languages to uproot extreme ideologies. Al-Azhar held an international conference recently to combat extremism in the Middle East and Africa.

The Cairo Center for Conflict Resolution and Peacekeeping in Africa is another of the Egyptian mechanisms by which it contributes to the African security framework. Cairo also links the African Union mechanisms with those of the Arab League.

The opportunity exists, through regional investment and careful negotiation, for Egypt and Sudan to see a significant growth in available water from both the Blue and White Niles over the coming decades, sufficient to increase the agricultural as well as the hydro-electrical base of the region. The construction of the "new Cairo" could also invigorate the Egyptian economy.

With regard to the sensitive strategic issue of Nile waters, Pres. Sisi in March 2015 took an active stance in helping structure the Declaration of Principles to pave the way ahead regarding the Grand Ethiopian Renaissance Dam.

Underpinning all of this will be the ability of Egypt to act as part of a stabilizing *bloc* on the region, something which is happening, *de facto*. The *bloc* of mutual interests, as discussed, includes Egypt, Israel, Jordan, Saudi Arabia, the UAE, Kuwait, and — as will become increasingly apparent — Ethiopia. There are major security as well as human and economic challenges to overcome, particularly to assist Saudi Arabia in stabilizing the Yemen situation before it ruins both Saudi Arabia and Yemen for decades to come. In all of this, Egypt may be able to persuade Saudi Arabia that its present efforts in Syria may be less than productive, and that the best way to stabilize and protect against Iranian influence would be through the creation of a broad new economic zone of progress which Iran could either welcome or ignore at its own peril.

Clearly, Egypt cannot be a beneficiary if Saudi Arabia and Iran engage in a protracted struggle. Egypt needs Saudi Arabia to be a stable littoral state on the Red Sea to ensure the growth of the Suez traffic. It also needs Saudi Arabia's investments in Egypt over the coming decade.

The tools which Egypt may well employ over the coming few years could well, eventually, include a strengthening of ties with Pakistan, also strategically placed on the Arabian Sea, and also now free — in a slightly different way than Egypt — to seek new solutions to its future. As well, Egypt is a major power in the regional intelligence field, and can use this capability to work with its neighbors in different ways.

There is a mutuality of interest between Egypt, Israel, Jordan, Ethiopia, Tunisia, some of the Libyan tribal leadership, Algeria, and Morocco, in countering extremist subversion. Other regional players, such as Kuwait and Oman, have a stake in this, as well, as do sub-Saharan states — particularly Kenya and Nigeria — which now find themselves facing the same enemies as Egypt.

This is a new world, already, and Egypt has the geography and resources to be central to it.

Some of the Saudi Arabian Army's 54 (two battalions) of Chinese-built PLZ-45 (Type 88) self-propelled 155mm howitzers prepare to move into Yemeni territory in 2015 to combat Yemeni Zaidi (Houthi) forces. This was the first combat use of the PLZ-45.

Turkish soldiers in an M-60A3 main battle tank and a 4x4 armored vehicle patrol on the road to the town of Beytussebab in the south-eastern, predominantly Kurdish Sirnak province, Turkey. The civil war between the Turkish Government and a range of Kurdish forces, mostly led by the PKK, is underway and is now intensifying.

VI

The Regional Impact of Fragility in Turkey and Saudi Arabia

Gregory R. Copley

TWO MAJOR REGIONAL STATES — three, if Iran is included — hold much of the key to whether the RedMed region will settle into stability or not in the coming decade: Turkey and Saudi Arabia. Both were, as 2016 dawned, embroiled in crises of an unprecedented scale. Iran, although a major factor in the RedMed region, posed challenges of a different nature to the RedMed.

Iran, unlike Turkey and Saudi Arabia, was not facing an imminent threat from internal pressures which could literally threaten its existence as a nation-state, nor its historical legacy as a distinct civilization. It had its own challenges, internally and in terms of its competitions with Turkey and Saudi Arabia, but the threats to Saudi and Turkish structural integrity were beginning to coalesce by late 2015.

Kurdish movements, largely centered around the restructured PKK (Kurdish Workers' Party) and other Kurdish groups, were, by late 2015, well embarked on a conscious, secret, and realistic process aimed at breaking up the Turkish State within the coming decade.[17]

17 See also: "Turkey Plunges Toward … What?", in *Defense & Foreign Affairs* ***Strategic Policy***, 5-6/2015.

At the same time, concerns were growing within Saudi Arabia that the Kingdom itself is facing systemic challenges which could see its break-up within a decade or two.

Related to the contextual framework, it is taken as a basis that Syria has already been largely dismembered. We await only the final details of how Syria would be shaped a decade hence.

Meanwhile, the most organized and imminent secessionist movement is that of the Kurds in Turkey.

There has been discussion of these prospects in the past, but what is now significant is that key regional players are gaming out future prospects in the event that the collapse of Turkey or Saudi Arabia, or both, becomes concrete. The Russian Government, in particular, is now aware of the very real potential for a break-up of Turkey along ethnic lines, something the Iranian Government has for some time been contemplating.

Significantly, however, Russia was not at this stage cooperating with the PKK in this endeavor, and had yet to determine whether a break-up of Turkey — or even the separate issue of a break-up of Saudi Arabia — favored Moscow's interests. It was significant, however, that Russian Ambassador to Ankara Andrey Karlov said pointedly on October 18, 2015, that neither the Kurdistan Workers' Party (PKK) nor the Democratic Union Party (PYD) were considered terrorist organizations by either Russia or the United Nations Security Council.

Already, seven states in the Kurdish east of Turkey have "declared autonomy" (as did several Kurdish provinces in northern Syria), although there is no indication at this time as to how effective or meaningful that declaration is. It is a topic not discussed in the Turkish media. State officials in this region have said that their territory is closed to Turkish entry. The Turkish Government response has been to escalate air strikes and, where possible, ground operations against PKK combatants in these areas.

PKK sources indicate that the group was now taking its campaign against Turkey to the main cities of the country, outside Kurdish areas.

PKK leader (officially the acting leader) Murat Karayılan (born in 1954), also known as Cemal,[18] has been active in planning the process, apparently from Kurdish northern Iraq, even though the Turkish Government has established "working relations" with the Kurdish Government of the region, under Masoud Barzani. But the importance of the liaison between the Kurdish fighting groups in Turkey and Syria with Kurdish political leadership under Remzi Kartal[19] in Brussels should not be under-estimated. There are some four-million Kurds in Western Europe, and Mr Kartal, 68, a dentist by training, is an important voice for them, too, in the halls of the European Union.

The Kurdish population of Turkey is officially estimated at just under 16 percent of the total population, although Kurdish groups claim that the percentage is as high as 22 percent; some statistics show it as high as 30 percent. And the Turkish Kurds are not the only minority chafing at the present governing arrangements. There are reportedly some three-million Kurds currently in Iraq, some 1.5-million of them reportedly from Turkey. There are a further two-million Kurds reported to be living in Syria.

And for the first time, Russian-Turkish relations are showing the strain of mutual frustrations. On October 6, 2015, Turkish Pres. Reçep Tayyip Erdoğan cautioned Russia

18 The Deputy Leader is reportedly Erdal Bahoj.

19 Remzi Kartal — who ran a private dental surgery in the city of Van until 1991 — became the executive member of the Kurdistan Parliament in Exile in 1995, and then the Kurdistan National Congress in 1999. He had been a member of the Turkish Assembly and had been a founding member of the first Kurdish party, HEP, in 1991. He is Co-Chair of *Kongra Gel*, the Kurdistan People's Congress (KGK), formally the new name for the PKK.

on its support for Syrian Pres. Bashar al-Assad, noting: "Our positive relationship with Russia is known. But if Russia loses a friend like Turkey, with whom it is having extensive cooperation, it will lose a lot. It has to know this." But Russia controls much of Turkey's fate, including the flow of oil and gas across from the Caspian Basin to the Black Sea basin and on to Europe.

Kurdish plans to dismember Turkey along ethnic lines — and all the Kurds are concerned about at present is getting an independent Kurdistan which would include at least the Kurdish regions of Turkey, Iraqi Kurdistan, and the Kurdish parts of Syria — would safeguard the westward flow of oil and gas from Iraqi and Iranian fields, as well as those from Azerbaijan, via Georgia, and into Kurdish areas of Turkey, egressing into the Mediterranean at Ceyhan. Kurdish sources insist that the ethnically Kurdish areas of Turkey go significantly beyond those shown by Ankara's maps, and now go up the Eastern regions of Turkey opposite Georgia, with frontage on the Black Sea.

What is being envisaged is the creation of a Kurdish state — with a population of between 25- and 30-million — which would embrace land in Turkey, Syria, and Iraq, but not yet Iran or Azerbaijan, which have Kurdish populations.

Certainly, Iran is concerned that the push for the creation of an independent Kurdistan, so long resisted by regional powers and the West, could also carve out Kurdish areas of north-western Iran. That has always been the issue which tempered Tehran's moves to use the Kurds to punish and break up Turkey. Now, it may be out of Iran's hands, and Tehran may have to make the best deal it can. The Iranian Kurdish guerilla group, Party for a Free Life in Kurdistan (PJAK), which began in 2004, is based in the Iranian provinces of West Azerbaijan, Kordestan, and Kermanshah. The

Iranian Government had effectively suppressed the group by September 2011, with PJAK groups moving across into northern Iraq for safe-haven with their related PKK colleagues. There were renewed clashes between Iranian Government forces and PJAK in 2013.

Kurdish sources, however, say that the PKK and its allies value Iranian cooperation, and believe that Iranian Kurdish areas would remain within Iran, and not form part of a new Kurdish state.

Meanwhile, the Syrian Kurdish People's Protection Units (YPG) are the main armed fighters of the Kurdish Supreme Committee, which acts as the Government of Syrian Kurdistan; the YPG has been preoccupied in the fight against DI'ISH (Islamic Caliphate, or Islamic State) in its area, and while Western sources indicate that its links to the PKK are questionable, PKK sources indicate a significant level of liaison.

And there is little doubt that the Kurds not only have momentum in their war against Turkey, they also have allies. Moreover, the Turkish imprisonment under a death sentence since 1999 of former (still nominal) PKK leader Abdullah Öcalan no longer has sway over the PKK movement, which has changed from its old marxist ideology, and is now far more inclusive as a political movement. Not that the Kurds are often united in anything other than their quest for a homeland.

Turkey went to the polls on November 1, 2015, following the hung parliament which was created by the June 7, 2015, elections. In those elections, the ruling AKP had sought sufficient majority to be able to change the Constitution without referendum, to allow for a transfer of executive power from the Prime Minister to the Presidency. The results, however, shattered the AKP's hopes, giving it the worst election results for some years. To survive in government, the AKP would have to enter a coalition. It won only 41 percent of the popular vote. The results of the June 7, 2015, election for the 550 seats in the Grand National Assembly were: 258 AKP; 132 CHP; 80 MHP; and 80 HDP. Initially, a chastened Pres. Erdoğan, and other AKP leaders, felt that they would have to agree to a coalition. No coalition was possible, however, so the outcome was to call new elections for November 1, 2015. In the interim, before those new elections, the only viable coalition partnering option had seemed to be the left-of-center, Kurdish-based HDP, with 80 seats. The two parties, however, disagreed on many issues, including electoral reform, so a coalition would have been delicately-based, and would have constrained the AKP from acting precipitously.

The HDP was, however, included in the interim Government which was only meant to hold until the November 1, 2015, polls. At that point, Pres. Erdoğan knew that he had to seriously degrade support for the HDP, so he ramped up the civil war against the Kurds. The President had hoped the revived attacks, including the bombing of an HDP rally in Ankara on October 10, 2015, would warn the Turkish people of the dangers of the PKK, would help bolster the anti-Kurdish vote in the new elections. It also served as a warning to the electorate to avoid the HDP.

Erdoğan's strategy worked, and HDP supporters stayed away from the new polls. The AKP was saved by the November 1, 2015, elections, which gave it more than 49 percent of the vote and 316 seats in parliament, more than the 276 seats needed for single-party rule but short of the 330 seats needed to take constitutional changes to referendum. There remained a significant proportion of the country which believed that the results were not credible, although there could be no doubt that the AKP had certainly mobilized its base.

Moreover, rather than a "vote for stability", which Pres. Erdoğan, had sought, Kurdish sources indicated that this would be the trigger for the start of the Kurdish revolt.

In response, there seems no doubt that in Turkish leadership circles, around Pres. Erdoğan, that the Government would attempt to invoke Article Five[20] of the Washington Treaty, which formed the North Atlantic Alliance (and therefore the North Atlantic Treaty Organization: NATO) if fighting against the Kurds came to a critical juncture. However, as an "internal attack", it is questionable as to whether any other NATO members would physically enter a war against the Kurds. Ankara, meanwhile, is attempting to build the case that the Kurdish operations are directed by an external movement. Certainly, Germany, in particular, has made it clear that neither DI'ISH (Islamic Caliphate/Islamic State) nor the Kurds — or an attempt to get NATO forces into a "buffer zone" to fight on the Syrian border — would be allowed to trigger an Article Five response to support Turkey.

Statements, such as the one made on October 6, 2015, to warn Russia, by Turkish Pres. Erdoğan highlight the sense in Ankara that Turkey is now under siege. It is, even after the November 1, 2015, election and the support which the Turkish Government blackmailed from Germany and other EU leaders.

More particularly, Mr Erdoğan's unconstitutional insis-

20 Article 5 of the Washington Treaty states: "The Parties agree that an armed attack against one or more of them in Europe or North America shall be considered an attack against them all and consequently they agree that, if such an armed attack occurs, each of them, in exercise of the right of individual or collective self-defense recognized by Article 51 of the Charter of the United Nations, will assist the Party or Parties so attacked by taking forthwith, individually and in concert with the other Parties, such action as it deems necessary, including the use of armed force, to restore and maintain the security of the North Atlantic area. Any such armed attack and all measures taken as a result thereof shall immediately be reported to the Security Council. Such measures shall be terminated when the Security Council has taken the measures necessary to restore and maintain international peace and security."

tence on acting as though he had the executive powers of the Turkish Government (a rôle delegated to the Head of Government, the Prime Minister) was geared toward attempting to recover the influence of his Muslim Brotherhood ideology party, the *Adalet ve Kalkinma Partisi* (Justice and Development Party, or AKP) for the November 2015 election. That worked, up to a point.

Kurdish sources indicated that their calls for secession would become evident and clear-cut within months, now that the November 1, 2015, parliamentary elections had confirmed that there would be no political solution to demands for the Kurdish territories to be given full autonomy or independence.

The sources also indicate that the Kurdish leadership was in touch with some of the other ethnic minorities in the country. Some 36 linguistic groups are found in Turkey, which is perhaps why Pres. Erdoğan chose to emphasize in December 2014 that Ottoman Turkish was now to be taught in all schools. "There are those who do not want this to be taught. This is a great danger. Whether they like it or not, the Ottoman language will be learnt and taught in this country," he said.

The widespread support for Turkey in Washington and London — based on 20-year-old stereotypical views of Turkey as the "great NATO ally" — may see a denial of the changing realities in the region, but the momentum for major change is now evident.

Meanwhile, sources in Moscow indicated that Russia would increase active support for the multinational Kurdish community, particularly those engaged in combat against DI'ISH, but there was no indication as to how far that support could go, or whether it would support Kurdish plans for the creation of a Kurdish state, largely carved out of Turkey, Iraq, and Syria. Kurdish sources indicated that their reluctance to talk directly with Russian officials or intelligence services stemmed from concerns over Moscow's deep ties with Ankara, even though a mutual animosity has traditionally existed between Turkey and Russia.

What remained of the Turkish-Russian dialog was, in any event, under severe stress by early October 2015. In any event, Moscow's long- standing relationship with Armenia tied in with the Kurds' long-standing relationship with Armenia, so the overlap of interests has almost certainly been in the consciousness of Moscow. Russia maintained its military presence in Armenia after the collapse of the USSR, keeping the 102nd Military Base of the Group of Russian Forces in Transcaucasia at Gyumi, Armenia, about 120km north of the capital, Yerevan. Some 3,000 Russian troops have been based there, along with a reported 74 main battle tanks, 17 infantry fighting vehicles (IFVs), 148 armored personnel carriers (APCs), 84 artillery units, 18 MiG-29 fighter aircraft, and several batteries of S-300 surface-to-air missiles.

Russian Pres. Vladimir Putin used his United Nations General Assembly speech on September 28, 2015, to show support for the Kurds. "We should finally acknowledge that no one but Pres. [Bashar al-] Assad's Armed Forces and Kurd militia are truly fighting the Islamic State," he said. "We should all remember what our past has taught us," Putin said. "We also remember certain episodes from the history of the Soviet Union. 'Social experiments' for export, attempts to push for changes within other countries based on ideological preferences, often led to tragic consequences and to degradation rather than progress. ... It seems, however, that far from learning from others' mistakes, everyone just keeps repeating them," he continued. "And so the export of revolutions, this time of so-called 'democratic' ones, continues."

However, the key to the evolving crisis in Syria is Russia's support for the 'Alevis/'Alawites.

After the so-called Russian violation of Turkish airspace (for nine seconds) by Su-30 and Su-24 combat aircraft on October 3, 2015, in the Yayladagi region of Turkey's southern Hatay province, and the Russian Air Force entrapment by Sukhois of two Turkish F-16s the next day, Prime Minister Ahmet Davutoğlu called the Russian Ambassador for a major berating. The Russian Ambassador listened to Davutolğu warning of the danger to Turkish-Russian relations, and then responded calmly that, as far as Moscow was concerned, no violation could have taken place because Russia did not recognize the Turkish annexation of Hatay Province in June 1939. France had no legal right to tear Hatay from Syria and hand it over to Turkey, so the predominantly 'Alawite/'Alevi Hatay should be returned to Syria.

Prime Minister Davutoğlu became outraged at the comment, and said that "these technicalities" were irrelevant. The Russian Ambassador responded that NATO's handling of the Baltic States on the basis of the US refusal to recognize the Soviet annexation in 1940 set a precedent for the relevance of pre-World War II annexations.

The bottom line, the sources said, was that Moscow was going to give Ankara a hard time over nationalities, particularly in support of the 'Alawites and their 'Alevi brethren in Turkey.

Significantly, the Turkish Government reported the confrontation over the airspace violation somewhat differently.

Prime Minister Davutoğlu said on October 5, 2015, that the violation was "a mistake". "Our Foreign Minister called the Russian Foreign Minister and said clearly that this should not be repeated," Dr Davutoğlu said in an interview with *Haberturk* TV.

"The information from Russia this morning was that this happened by mistake and that they respect Turkish borders and that it will not be repeated. Whoever violates our airspace, our rules of engagement are clear. ... We will warn whoever violates our borders and our airspace in a friendly manner. This country can be whichever and Russia is our neighbor, our friend. In this way, there is no tension between Turkey and Russia. The Syria issue is not a Turkish-Russian crisis," he said.

Turkey was, however, moving to make the issue a NATO one, to invoke Article Five of the Washington Treaty. NATO Secretary-General Jens Stoltenberg said on October 5, 2015, that the airspace violation was "unacceptable", although failing to comment on the consistent NATO violation of Syrian airspace. Mr Stoltenberg said that he had met with Turkish Foreign Minister Feridun Sinirlioğlu on October 5, 2015, to discuss Russia's moves. "Russia's actions are not contributing to the security and stability of the region," he said in a statement. "I call on Russia to fully respect NATO airspace and to avoid escalating tensions with the Alliance."

The matter escalated substantially on November 24, 2015, with the shooting down by a Turkish Air Force F-16 of a Russian Air Force Sukhoi Su-24 strike aircraft which may (or may not) have briefly violated Turkish airspace.[21]

These incidents directly impact the RedMed region, with Moscow moving strenuously into a program of economic and indirect sanctions and punishment of Turkey. Ankara's response has included — with the support of the US Obama Administration — a tenuous and artificial *rapprochement* with Israel, in order to obtain access to Israel's

21 For greater detail on the incident and the broader strategic ramifications, see Copley, Gregory R.: "Russia and Turkey: Not War in the Offing, But Something Far More Important" in *Defense & Foreign Affairs **Strategic Policy***, 11-12/2015.

offshore gas production to compensate for the imminent loss of energy supplies which Moscow had the ability to cut.

Significantly, with Turkey now escalating the confrontation with Russia, Moscow has little disincentive from supporting — directly or indirectly — the Kurdish thrust for a separate state, particularly given the knowledge of Kurdish plans to sustain a stable pipeline framework through the proposed new Kurdish state from the Caspian, Iraqi, and Iranian regions through to the Mediterranean at Ceyhan, in Turkey's heavily Arab-Kurdish Adana province, near the capital Adana (fifth largest city in Turkey), which is also in proximity to the Turkish Air Force/NATO air base at Incirlik.

The Turkish opposition newspaper, *Today's Zaman*, carried a report on its website on December 17, 2015, noting that "Turkey comes close to the brink of civil war", and confirmed that "Around 10,000 troops backed by tanks have launched fresh massive operations in various cities in the region to clear the area of Kurdistan Workers' Party (PKK) terrorists. The fighting between the security forces and the youth branch of the PKK has been ongoing in urban areas since its resumption in mid-July [2015], when the Government declared that it had frozen the Turkish Kurdish peace process, which lasted only two-and-a-half years. The process was intended to end the country's almost three-decades-old armed uprising that has taken the lives of around 50,000 people, including members of the security forces, through non-military means."

The report continued: "Since the collapse of the peace process and its replacement by heavy fighting since mid-July [2015], around 300,000 people have been forced to leave their homes due to clashes and curfews, while schools, government buildings and houses have been destroyed and infrastructure crippled."

Ankara continued to attempt to suppress all news of internal conflict, and re-doubled its efforts to bring about success for its operation to overthrow the Syrian Government of Pres. Bashar al-Assad.

By November 2015, the head of Turkish intelligence had begun calling for the recognition of DI'ISH as a nation-state, with a diplomatic mission in Ankara. Moscow noted the gesture to begin the creation of the Islamic State as a normal sovereign entity. Ankara had learned from Washington's creation of a sovereign Kosovo. But who else noticed?

Saudi Arabia

THE THREAT OF STATE collapse and break-up is not confined to Turkey. In Saudi Arabia, a major Royal prince has in late 2015 cautioned his family about the dangers facing the Kingdom.

On September 22, 2015, *Middle East Eye* broke details of a letter circulated among Saudi princes from a grandson of the late King 'Abd al-'Aziz al-Sa'ud, which blamed incumbent (and ailing) King Salman bin 'Abd al-'Aziz al Sa'ud for creating unprecedented problems which endangered the monarchy's survival. "We will not be able to stop the draining of money, the political adolescence, and the military risks unless we change the methods of decisionmaking, even if that implied changing the king himself," he said.

The threats to Saudi cohesion have been evident for some time, but have been contained by a fairly efficient internal security capability. It was in early 2009 that Iranian leaders decided to clandestinely support the Shi'a revolt in eastern regions of the Kingdom, where a "Republic of Eastern Arabia" had been declared on May 15, 2009, by a group reportedly led by *Sheikh* Baqir al-Namr.[22]

22 See, Bodansky, Yossef: "Iran Moves at Highest Level to Support the Newly-De-

Today, however, Saudi Arabia's cohesion and prospects are even more questioned because of its commitment to the new adventures in Yemen and the ongoing one in Syria, under the new leadership slate which followed the death of King 'Abdallah bin 'Abd al-'Aziz al Sa'ud, who died on January 23, 2015, and the removal of his chosen nominee for Deputy Prime Minister and Crown Prince, Muqrin bin Abd al-'Aziz, on April 29, 2015.[23] Since then, the economic prospects for the Kingdom have continued to decline: largely because of the view of protracted low oil prices; and partly because the Kingdom had over-extended itself in the war against the Houthi-led force which had taken over much of Yemen; and in the war against Syrian Pres. Bashar al-Assad.

Meanwhile, a major study in the *Journal of Petroleum Science and Engineering*, published in April 2015, anticipated that Saudi Arabia would experience a peak in its oil production in 2028, followed by inexorable decline.[24]

The Kingdom's immediate cash requirement caused it to sell bonds over the Summer of 2015 to raise at least $4-billion, the first time that Saudi Arabia had gone to the bond markets for eight years. It also gave signs that it would be withdrawing funds from global asset management firms, and had removed $50- to $70-billion over the six months to September 2015. It was estimated that Saudi Arabia's current account deficit would reach 7.5 percent of GDP. This represented a major turnaround in the Kingdom's fortunes. King Salman bin 'Abd al-'Aziz al Sa'ud has ramped up spending, particularly on defense, since taking over

clared 'Republic of Eastern Arabia'." In *Defense & Foreign Affairs Special Analysis*, May 18, 2009; and in *Defense & Foreign Affairs* ***Strategic Policy***, 4-2009.

23 See: "The Strategic Impact of the Saudi Leadership War and the Retreat to Conservatism", in *Defense & Foreign Affairs Special Analysis*, April 30, 2015. Also in *Defense & Foreign Affairs* ***Strategic Policy***, 4-2015.

24 See: "Forecasting OPEC crude oil production using a variant Multicyclic Hubbert Model", by Mohsen Ebrahimi and Nahid Cheshme Ghasabani, in the *Journal of Petroleum Science and Engineering*, published by Elsevier, online as of April 20, 2015.

from King 'Abdallah.

One estimate noted: "Saudi state revenues, 80 percent of which come from oil sales, are heading downwards, terminally", and oil exports would reach zero within two decades. Population numbers, however, were on the upswing, from 29-million (current estimate) to 37-million by 2030.

The Saudi Arabian Government released its 2016 budget on December 28, 2015, along with details of actual 2015 budget performance. The 2015 budget deficit, as it transpired, had been $97.8-billion, on expenditures of 975-billion *riyals* ($260-billion), 13 percent above target. Economy Minister Adel Fakeih said that 20-billion *riyals* ($5.33-billion) of the 2015 spending was due to increased military and related spending in operations against Yemeni Zaidi forces. The Government planned to cut overall spending in FY 2016 to 840-billion *riyals*, some 25 percent ($55.94-billion) of which would be allocated to defense and security budgets. Healthcare would consume 35 percent of the budget. The deficit was expected to narrow to 326.2-billion *riyals* ($87-billion) in 2016, from 367-billion *riyals* in 2015. The 2015 budget deficit represented about 16 percent of GDP. The Saudi Central Bank's foreign asset reserves fell more than $95-billion in the first 11 months of 2015 to $627.7-billion.

Several factors apply: firstly, oil prices will eventually rise, although perhaps not sufficient to cover the decimation wrought on the Saudi foreign exchange reserves; secondly, operational spending on the Yemeni war would likely exceed, in 2016, the costs in 2015, making the budget deficit worse than anticipated for the year. And the population climb in the Kingdom — in which immigrant groups will continue to increase substantially over Saudis — may taper off, but the social unrest should be expected to become significant in the coming decade.

The September 24, 2015, stampede among pilgrims in Mina, Mecca, causing at least 1,400 deaths, was politically damaging, too, to the new Salman Administration.

The war against Yemen is taxing Saudi Arabia in many ways, as well, and causing anxiety among other regional states. The Kingdom's relationship with Djibouti and Ethiopia — as described in Chapter III — has become critically damaged, and its relationship with neighboring Oman, always "careful", worsened substantially as a result of differences over the Yemen war. Oman, in fact, had, during 2015, allowed Iranian Air Force overflights to deliver weapons and logistical support for the Houthi Zaidis in Yemen.

But even Qatar, part of the Saudi coalition against the Houthis (apart from airpower, it has despatched 1,000 troops to fight in Yemen), was in 2015 conducting contingency planning for the possibility of civil war in Saudi Arabia in the event of an economic meltdown there, accompanied by social unrest involving a range of sects, tribes, and migrant groups. That would open profound risks as well as opportunities for Qatar, which has claims on large tracts of Qatari-claimed lands occupied by the Kingdom. As one Qatari military analyst privately noted: "A civil war in a society armed with advanced weapons would have devastating consequences for the region as a whole."

So the prospect exists for a very new and diverse geopolitical landscape over the coming two decades in the Eastern Mediterranean and Arabian Peninsula. The US and Europe would probably fight politically to preserve the *status quo*, but the momentum may be against them.

Russia, Iran, and the People's Republic of China may accept the changes and attempt to benefit from them. Or perhaps not. Transitions are rarely peaceful.

Indeed, Saudi Arabia's moves at the beginning of 2016 ensured that the coming year would be interesting.

Saudi Arabia Moves Toward Confrontation

Saudi Arabia's Sudeiri-line leadership of the House of Sa'ud began 2016 by a major push to save its position and control of the Kingdom. It was also a bid to solidify regional power as the Kingdom moved well beyond the shadow of the major power relationships which had dominated its existence since the creation of the State in 1932.

Saudi Arabia's execution of Shi'a cleric *Sheikh* Nimr Baqr al-Nimr, 56, on January 2, 2016, along with 46 other political dissidents also sparked a divide between Iran on the one hand and Saudi Arabia and several of its allies on the other.[25] It was a salvo in a geopolitical war, which is far more deeply-constructed than the seemingly intra-Muslim sectarian war in which the surface noise — or carrier wave — is characterized by symbolic actions and rallying calls.

That is not to say that there was no religious element — even motivation — to the actions. There clearly was a significant religious aspect, but the actions of the Saudi leaders stemmed mostly from an attempt to preserve their position at the head of the geopolitical entity of the Kingdom as well as through their claim to religious leadership from the position which they seized in 1932 as Custodian of the Two Holy Mosques.

This was a crisis which had been brewing for some months, and the execution merely served — and deliber-

25 Saudi Arabia on January 2, 2016, executed — shot or beheaded across 12 sites — 47 dissidents, including *Sheikh* Nimr Baqr al-Nimr, a mid-level (but highly significant) Shi'a Saudi cleric from Al Awwamiyah, a village in the predominantly Shi'a Eastern Province of Saudi Arabia. The sudden concurrent, multi-site series of executions triggered several days of public protests in several Iranian and other Middle Eastern cities in response, and the Saudi Embassy in Tehran was firebombed on January 2, 2016, as well as the Saudi consulate in Mashhad. Saudi Arabia then broke diplomatic relations with Iran, and Bahrain and Sudan followed suit, while the United Arab Emirates, a major trader with Iran, downgraded diplomatic relations with Iran, and Kuwait then followed by withdrawing its Ambassador to Tehran. The Saudi executions drew widespread international condemnation, but the Iranian Government was also criticized for not stopping the protests which destroyed the Saudi legations.

ately so on Riyadh's part — as a catalyst for more visible divisions. *Defense & Foreign Affairs* Senior Editor Yossef Bodansky said that the execution potentially cast *Sheikh* Nimr in that catalytic position as "the Archduke Ferdinand of the Middle East". As Bodansky also reported on May 18, 2009: "The fiery *Sheikh* Nimr Baqr al-Nimr (40+) of Awwamiyah is reported to be the spiritual guide of the new Republic [of Eastern Arabia]" which was declared in the Shi'a areas of Saudi Arabia, reason enough for the Kingdom to want his head: he had called for the break-up of the Kingdom.[26] *Sheikh* Nimr was not the unimportant cleric many Western media reports claimed.

But, by the beginning of 2016, the Saudi leadership needed to galvanize support for its leadership, even its legitimacy, both within the Kingdom and abroad. Concerns over internal social and political fracturing may, arguably, have been more important than winning additional support from other Sunni states for Riyadh's primacy in wars against Yemen and Syria.

The ramifications of the symbolism of the executions — which immediately appeared profound within the region — seemed likely to be even more significant than the reactive outcry indicated.

Of significance is that the event once again, as a byproduct, severed US-Saudi Arabian ties, or served to show that Riyadh had moved beyond the influence of Washington.

US Secretary of State John Kerry had, immediately after the anti-Saudi riots which destroyed Saudi diplomatic missions in Iran, telephoned Iranian Foreign Minister Mohammad Javad Zarif to discuss the issue. He then telephoned his Saudi counterpart, Foreign Minister Adel al-Jubeir, only to be told that the Saudis did not wish to speak

26 See, Bodansky, Yossef: "Iran Moves at Highest Level to Support the Newly-Declared 'Republic of Eastern Arabia'", in *Defense & Foreign Affairs Special Analysis*, May 18, 2009.

to him. This was the worst snub to the US by Saudi Arabia since the late King 'Abdallah bin 'Abd al-'Aziz al Sa'ud cut short US Pres. Barack Obama's State visit to the Kingdom on March 28, 2014.

Thus, the growing rift between the Saudi *bloc* and Iran now appears to be beyond the scope of the US to influence. This begs the question as to how the Saudi-Iran confrontation would now progress, and who would benefit or suffer from it. It was immediately clear that the event threw into chaos negotiations for a peaceful end to the Saudi-led war in Yemen. Continuation of that conflict threatens to exacerbate the Saudi budget crisis, which has already led senior Saudi princes to challenge the wisdom of the current leadership.

This may cost Saudi Arabia dearly, particularly if the US position as a Saudi ally is simultaneously thrown into doubt. What now seems significant is that the only "major power" capable of taking advantage of the situation — given the overall spread of conflict from Syria to Yemen, and even across the Red Sea — is Turkey. The US and European Union (EU) have no influence; Russia is perceived as being in support of Iran; and the People's Republic of China (PRC) is attempting to take no side in the rift.

It also begs the question as to what provoked the Kingdom into the mass executions of dissidents as the "unified symbolic message" which it was.

Part of it lay in the fact that the Saudi leaders had come under increasing pressure from domestic and regional hard-line Wahhabists to demonstrate their credentials and legitimacy. This included pressure from Islamic State (DI'ISH) supporters as well as domestic radicals and even mainline members of the Royal Family who had become disillusioned with the rash destruction of the Saudi economy by escalating spending on foreign wars.

It is significant that, of the 47 dissidents executed on January 2, 2016, three were indeed Saudi Shi'as, but about a third were *al-Qaida*-linked Saudis, the majority were Sunni.

In the meantime, although the Gulf Cooperation Council (GCC) called for an emergency meeting to discuss the crisis, it was clear that the GCC itself was not in harmony, and nor could it represent a balanced view of the situation. Iran, for example, is not represented in the GCC. But within the GCC there is strong antagonism between Saudi Arabia and the Sultanate of Oman because the Saudis felt that Oman had acquiesced to Iranian delivery of weapons and combatants to Yemen, overflying Omani airspace to skirt Saudi Arabia and possible US Navy interdiction. Those flights — by Iranian Air Force Ilyushin Il-76MD and Boeing 747 transports — had transported major weapons systems, such as the *Iskander-E* (NATO codename SS-26) battlefield ballistic missile, and *HizbAllah* combatants to the Zaidi Shi'a Houthi forces fighting the Saudi coalition.

Oman had, in November 2015 (and later), played a rôle in bringing the Houthis into the Geneva Peace Talks on Yemen. Talks between the Yemeni Government of Pres. Abd al-Rab Mansour al-Hadi and the Houthis began in Geneva, under United Nations auspices, in December 2015 and on December 15, 2015, both sides called for an immediate ceasefire for seven days. [A previous ceasefire, in June 2015, collapsed before the parties met.] Breaches of the ceasefire began almost immediately, and the talks again broke down and ended by December 20, 2015.

Significantly, the Saudi-led coalition fighting in Yemen had been addressing more than merely the Zaidi Houthis; various Sunni *al-Qaida* groups were also in conflict with the coalition. On January 1, 2016, pro-Hadi Government fighters killed *al-Qaida* judge Ali Abed al-Rab bin Talab (aka Abu Anwar) and three others in Abyan province of southern Yemen. Abu Anwar was the chief *al-Qaida* judge in the neighboring Hadramaut region, which also abuts the southern Omani border.

The Saudi coalition forces were making a major push in late December 2015 against Houthi forces in areas just outside the capital, Sana'a, and the coalition was pushing additional forces into the area. Qatar, at odds with Saudi Arabia on several key issues, was supportive of the Kingdom on this front. In September 2015, Qatar committed 1,000 troops, supported by 200 armored vehicles and 30 Boeing AH-64 *Apache* helicopter gunships, to support the Saudi operation. More Qatari forces were expected to follow that to help the coalition take the Jawf Governorate of Yemen. Coalition casualties on the ground have been heavy, and several combat aircraft have been lost to surface-to-air missile fire and accidents: the Royal Moroccan Air Force lost one F-16 in May 2015, and the Bahraini Air Force lost an F-16 in Jazan region, near the Saudi-Yemen border "due to a technical error" in December 2015.

Broader Ramifications: The polarizing of what was portrayed as a "Sunni-Shi'a rift" — in reality a geostrategic rivalry — had a range of other attendant issues:

1. Domestic Security: The attempt by the Saudi leadership, acting in the name of King Salman bin 'Abd al-'Aziz al Sa'ud, but particularly galvanizing around his son, Defense Minister and Deputy Crown Prince Mohammed bin Salman, 30, to suppress the growing dissent within the extended Royal Family over the conduct of national affairs and the growing decline in the economy and economic outlook. This known dissidence has reinforced other, non-Royal, dissidents, and all opposition was now being treated as "terrorism".

2. Regional Fears: There is no doubt that there has been concern within the Government, as well as regionally, that the Russian intervention in the Syrian conflict has taken the initiative away from the collection of often mutually-opposing "Sunni forces", and restored the initiative to the Government of Bashar al-Assad and, therefore, to his ally, Iran. Moreover, successful actions by the Shi'a-dominated Iraqi Government against Islamic State (DI'ISH) forces in recent months had also reduced the impetus of Sunni forces in Iraq, which could be influenced by Saudi Arabia, and given momentum to Iran. Iran, too, was gaining in strength by its ability to shape the conduct of the war in Yemen, by ensuring no quick victory for Saudi-led forces. This was compounded by the reality that Iran would only gain in economic and military strength (and regional influence) as the international sanctions against it were lessened as a result of the 2015 deal which Iran struck with the G5+1 team, led by the US, ostensibly to end Iranian nuclear weapons programs. Little wonder that the Saudi leadership viewed the US as no friend.

3. Economic Pressures: Both Iran (freed from oil sanctions) and Saudi Arabia committed to increasing crude oil production in early 2016. Saudi Arabia was counting on the fact that falling international oil prices were helping to drive US shale oil producers out of business. Although the US shale production technologies have proven to have evolved more rapidly towards low-cost viability than expected, the decline in oil prices by more than 60 percent since June 2014 has driven the number of US oil rigs from a peak of 1,609 rigs in October 2014 to a low of 524 on December 11, 2015. Some Saudi analysts would argue that the gamble to continue strong oil production in the face of low prices was paying off. But the continued high budget deficits which the Saudi Government is incurring threaten to exhaust its $700-billion of financial reserves within five or so years. And Saudi Arabia's oil and gas reserves have been estimated to last — at the current rates of depletion — for only another 18 or so years. Saudi Arabia's policies, however, are a gamble that the low price of oil has been determined largely by the increased supplies made available by US shale exploitation. This is not the case: current and anticipated global economic malaise has also limited demand. Moreover, a rise in global economic fortunes (and therefore energy demand) would also trigger a resumption of the rise in US production, quite apart from the reality that a future US administration would at some stage approve a widening of the exploitation of the larger US oil and gas reserves which currently have been ruled out of bounds. But for the Saudi leadership the existential nature of short-term survival outweighs long-term considerations.

4. Rash Actions: Saudi Arabia's very real concerns that the Yemen war was dragging on far longer than it could realistically sustain has made it undertake rash actions, such as the steps, as noted elsewhere in this book, in concert with the United Arab Emirates, in 2015 to cut relations with Djibouti and bolster Eritrea and potentially Somaliland. The way in which this occurred *de facto* caused a threat to Ethiopia, which depends vitally on Djibouti for its exports. Despite Saudi Arabia's feelings of distrust for the US at present, the US had, in fact, supported Saudi Arabia and the UAE over its rift with Djibouti, at potentially significant strategic cost to the United States' ability to sustain power projection in the Red Sea. This, too, jeopardizes Ethiopia's security, and the polarization of Saudi-Ethiopian feelings seems likely to impact on Saudi Arabia's and the UAE's military support for Eritrea, reviving the belief by Pres. Isayas Afewerke in Asmara that he could (with help from Riyadh and Abu Dhabi) resume his proxy war against Ethiopia.

It was not insignificant that Ethiopia was jarred in late

2015 to learn that the US had unilaterally withdrawn from its use of the Ethiopian Air Base at Arba Minch, from where it had conducted UAV strike and reconnaissance operations against Somalia's *al-Shabaab* and other *jihadist* rebel groups. The Ethiopian leadership had pinned all its strategic hopes on the protective umbrella of the US, but that was now being gradually eroded. Rash Saudi actions have continued elsewhere, and the least obvious of these, the construction of Wahhabist mosques around the world, may finally face opposition, not just in Western countries, but even in Pakistan and Ethiopia. But Saudi Arabia's financial support for anti-Chinese *jihadist* groups, such as the "East Turkistan Independence Movement" has alienated PRC support for Saudi Arabia. Russia, by September 2015, had become the primary oil supplier to the PRC, a situation which seemed likely to continue. There is little doubt that, as US/Western strategies persist, and Saudi-Qatari-Turkish support for *jihadism* continues, Russia and the PRC have found a growing identity of strategic interests.

5. Egypt and Others: Where does the Saudi direction leave Egypt? Saudi Arabia and the UAE were vital economic supporters of Egypt when popular dissent drove then-Pres. Mohammed Morsi from office in mid-2013. They remain the most critical investment partners of incumbent Pres. Abdul Fatah al-Sisi. As a result, Egypt has been compelled to support the Saudi-UAE drive into Yemen, and to remain silent on Saudi actions — in concert, often, with Qatar and Turkey — in Syria, even though Cairo has grave concerns about Saudi actions. The Egyptian Government has also remained notably quiet on issues such as the Saudi and UAE rift with Djibouti and Riyadh's and Abu Dhabi's move to create military bases into (and provide military and economic support for) Eritrea. Saudi Arabia's and the UAE's actions with regard to Yemen, Eritrea, and Djibouti profoundly affect Egypt's most vital trade route: the Suez/Red Sea sea-line of communication (SLOC). But Saudi Arabia's seemingly messianic war against Shi'ism in Yemen, Syria, Iraq, and domestically within the Kingdom has made some of its historical allies, such as Pakistan (where some 20 percent of the population is Shi'a), nervous.

Not surprisingly, Pakistan's National Assembly voted unanimously on April 10, 2015, to reject Saudi Arabia's request to join the anti-Houthi military coalition, despite the extremely close ties which Pakistan Prime Minister Nawaz Sharif has had with the Kingdom. Even for Egypt, which has no Shi'a population of any significance, Saudi Arabia's marriage with intra-Sunni rivals Turkey and Qatar causes concern. Clearly, Egypt is seeking cultural and strategic leadership within the Arabic-speaking world, but has not sought the kind of religious leadership of the Sunni community which has been sought by the Saudis, Qatar, and Turkey. As if to reinforce his secular credentials over the religious, Egyptian Pres. al-Sisi on December 26, 2015, met with Iraqi citizen Nadia Murad, a Yazidi woman who had been kidnapped and assaulted by DI'ISH fighters when they took control of Sinjar city in August 2014. The Yazidi religion of Sharfadin is pre-Islamic and linked to ancient Mesopotamian religions.

Unspoken in all media reporting is the position of Israel, which has significant, close intelligence ties with Saudi Arabia, but which is concerned over Turkish, Qatari, and Saudi (and US) attempts to replace Syrian Pres. Bashar al-Assad. The latest Saudi moves seem likely to consolidate consultations between Egypt, Israel, and Jordan, in particular.

Bahrain, committed to the Saudi coalition in Yemen, and yet with a majority Shi'a population, immediately sided with Saudi Arabia in breaking diplomatic relations with Iran. This may have been an attempt to limit any Iranian

Embassy-based inspiration of riots against the Government, but even so, Iran's reach into the Bahraini community is clearly deep and discreet, and Bahrain could soon experience significant unrest. This would impact US and British military deployments based in that Kingdom.

6. Syria and DI'ISH: The question remains as to whether the significant upsurge in internal security actions, which the January 2, 2016, executions symbolized, would detract from Saudi Arabia's physical ability to fund and staff military, intelligence, and *jihad*-related activities against Syrian Pres. Bashar al-Assad and act in tacit support of DI'ISH (Islamic State). Saudi Arabia cannot afford to absent itself from the leadership of the Sunni cause in Syria and Iraq, but neither can it afford the physical cost of a significant military contribution. Thus, the Syria/Iraq intervention would likely continue to be a function of the intelligence services, particularly the General Intelligence Presidency (*Al Mukhabarat Al A'amah*) and the military intelligence agencies.

In the short term, Saudi Arabia seems likely to continue to escalate its military activities in Yemen to attempt to force a settlement — and a public indication of victory — as quickly as possible. Iran's function would be to defer, mitigate, or defeat this goal. Saudi Arabia's next step, already underway, must be to mobilize to the maximum the radical and moderate Sunni Muslim support, galvanized around Wahhabism, to fight its geopolitical war against Iran as a religious war.

In such a war, assertions of moral and dialectical correctness are major weapons of legitimacy, but they are not the underlying cause of the conflict. But by launching a new dynamic, Saudi Arabia has helped further unleash proxy and follow-on (or competing non-governmental militancies) over which it has little or no control. It is true that the

Saudi leadership has not thought about second- or third-order effects of its current, expedient actions. But it is also true that the factors actually beyond measurable calculation or control, such as the social motivations within the Muslim world's various factions, may be the determinants of whether the situation could trigger an irreversible and major conflict, or the collapse of states.

Saudi Arabia could still be one of the major casualties of its own actions. And while the break-up of Yemen, once again, now seems like a foregone conclusion, the break-up of Saudi Arabia could occur with equal speed.

McDermo
& F

VII

The Regional and Extra-Regional Impact of the Refugee Crises

Ambassador Princeton N. Lyman

THIS REGION IS TORN BY several conflicts that threaten the stability and unity of the region and the interests of the broader international community. Refugees are an important part of this scene.

There are more than three-million refugees currently in this region, and hundreds of thousands more from the region have fled to Europe, the Middle East and elsewhere. In some cases like in Kenya, refugees have been residing in neighboring countries for decades. In other cases they are recently arrived. In several cases, moreover, they add to the millions of internally displaced in these same countries. [See the chart on the next page.]

Under any circumstances, three-million refugees crossing or sometimes crisscrossing borders, putting demands on host countries and the international community and often linking back to the conflicts from which they have fled, cannot help but have far-reaching political, economic, and cultural impacts.

Sources and Hosting Countries for Refugees

Many of the countries in this region are both the source

and the host to refugees reflecting the volatility of the region. Many of them also have large numbers of internally displaced persons (IDPs) adding to the turmoil of the situation.

	Refugees From*	Refugees In*
Egypt	25,000	236,000
Eritrea	363,000	
Ethiopia	86,000	659,000
Chad	49,000	453,000
Kenya	551,000	
Somalia	1,106,000	
South Sudan	616,000	248,000 (plus 1.3-million IDPs)
Sudan	665,000	277,000
Uganda	385,000	
Yemen	2,600	264,000 (plus 2.3-million IDPs)

*UNHCR (figures rounded)

Political Implications

Refugees are often portrayed in humanitarian terms, which is valid. They are fleeing conflict and often arrive destitute and broken. But they are not necessarily political eunuchs. They often come from one side or the other in a neighboring conflict. Once settled, whether in towns or in camps, they may provide supplies and funds back to their colleagues in the fight. Refugee camps sometimes serve as rest and recuperation and recruiting stations for the fighters. How much the host country polices the refugees, prevents on the one hand, or looks the other way about their support for the fighting across the border, bears on its relationship with the country from which the refugees come.

There are many examples. Sudan once bombed the Yida refugee camp in South Sudan, claiming that SPLM-N soldiers from the Sudanese province of Southern Kordofan frequented the camp and that some of the supplies meant for the refugees were being sent across the border. Refugees in Ethiopia from the Sudanese province of Blue Nile, partners of the SPLM-N fighting the Sudanese government, pose a sensitive problem for Ethiopia — though sometimes perhaps leverage — in its relations with Sudan. South Sudanese refugees pouring across the border into the Gambella region of Ethiopia represent a different issue. Some 95 percent of these refugees are women and children, an unusually high percentage. This suggests that the men, voluntarily or otherwise, are staying to fight while the host country and the international community bear the costs of supporting their families.

Somali refugees in Kenya, some 300,000 in number, pose a special problem. They occupy the largest refugee camp in the world just over the border from Somalia. Many have been there for decades and posed only minor problems. Once Kenya entered the war against *al-Shabaab* in Somalia, however, that organization carried out terrorist attacks in Kenya; eg: the attack on the Westgate Mall in 2013 and on a school in a Kenya Army compound in 2015. Kenya worries that both the Somali refugees and the large number of Somali-Kenyans may be providing pathways and even support for the terrorists. Kenya's sometimes harsh response, however, has alienated the Somali-Kenyans in particular and may be aggravating the threat.

Sudan now faces a particular problem. When South Sudan became independent in 2011, Sudan stripped both citizenship and most work permits from the hundreds of thousands South Sudanese living in Sudan. It did not want a *bloc* of voters, or even residents, whose loyalties may not be to the Sudan Government, and especially perhaps sympathetic to the opposition party, the SPLM-N. Hundreds of thousands thus went south to the newly independent South Sudan. But since civil war broke out in South Sudan, some 200,000 South Sudanese refugees have crossed back into Sudan. Some want to return to their previous homes and jobs in Khartoum and elsewhere. Sudan has allowed the refugees to stay for now but it has not decided their long term fate. This is a problem, and again complicates relations between the two countries.

Farther from the scene, Eritrean refugees crossing the Sinai have been the victim of ruthless gangs. Once in Israel, they, along with South Sudanese refugees, have posed a political problem for that country, which is worried about any long term impact on the religious demography of the country, and raising instances of xenophobia. Recently an Eritrean refugee was stomped to death by Israelis when he was mistakenly suspected of being a Palestinian terrorist. Israel has sought to return refuges to their countries of origin, eg: South Sudan, even when the situation is not conducive and despite objections of UNHCR.

Economic and Environmental Costs; Lost Opportunities

While UNHCR and NGOs provide basic support for refugees, that support is sometimes minimal and refugees must find resources or work in the host country. In many cases, they have been accused of deforestation as they seek wood for fuel. Without proper sanitation facilities, they despoil the environment. As workers they compete with local labor. When they are provided schools, health services and other amenities that exceed those for the local population, there is resentment. Finally, there are security concerns, such as attacks on women in the camps or when they are out seeking fuel, beyond what UNHCR can manage, and for which the host country bears the principal responsibility.

The costs to the international community are also great. UNHCR's worldwide budget for 2015 was $7-billion but it believed this would cover little more than 60 percent of its requirements. For East Africa and the Horn it was $1.5-billion (and rising) in 2015. Bilateral — very largely the US — and regional (eg: the EU) budgets are devoted to these needs as well. The real cost of these conflicts go well beyond direct refugee assistance, however. There are large expenses for peacekeeping, IDPs, food assistance in war-torn areas, and other expenses. There are currently three UN peacekeeping missions in Sudan and South Sudan, and an Africa Union operation in Somalia. Even before the civil war erupted in South Sudan in 2013, the international community was feeding a third of South Sudan's population because the agriculture in the country had never recovered from the long civil war with Sudan.

When I became the US Presidential Envoy for Sudan and South Sudan in 2011, I was asked by Congress how much the US had spent on the peace process for those two countries since the Comprehensive Peace Agreement of 2005. Adding in peacekeeping, emergency assistance, and a relatively smaller amount for development, the figure was $10-billion, a figure that astounded the Congress as well as me. By the end of 2015 that figure had grown to probably $15-billion. Adding in the contributions of other countries and international organizations, the total figure could approach $30-billion, just for those two countries.

Those are Marshall Plan numbers. Just think what $30-billion could have done in this region without conflict. Think of the water projects, the roads which could have been built, the schools and teachers provided, the industrial infrastructure created. These are the forgone opportunity costs of conflict in this region.

Wider Repercussions

There has long been an understanding between the international community and Africa about refugees. Legally, refugees are defined as those who have a legitimate fear of persecution if forced to return to their home country, distinguishing them from "economic migrants". In Europe and the US these determinations have been historically made individually before someone is given asylum. But in Africa, torn by conflicts sending hundreds of thousands of people across borders, individual determinations have been impossible in those circumstances, while closing the borders or forcing these people back would have been reprehensible. So Africa determined that it would welcome these people, and treat them all as refugees. In return, UNHCR and the donor community would recognize them as such as well and provide for their needs. This allowed western nations to keep massive numbers from their borders and to continue to process potential asylum seekers for resettlement in their countries through individual interviews.

That understanding has now broken down. Primarily from the rush of Syrians but also featuring Africans from Eritrea, Libya, Somalia and elsewhere, hundreds of thousands of refugees are now pouring into Europe. By the end of October 2015, more than 700,000 had arrived in Europe passing through one country after another to find asylum and opportunity. This was already more than three times the number of refugees resettled in Europe in all of 2014. The continent is seemingly overwhelmed as it seeks to respond to this crisis.

The situation has created tremendous strains within the European Union, perhaps the greatest threat to its unity since its creation. Several members have refused to accept the EU Commission's proposed quota system for resettling this massive influx. Some countries have put up walls to redirect the flow to other member countries, despite EU objections. There has been political backlash in countries which have welcomed the refugees as well as in those who are resistant. The refugee crisis has strengthened domestic support for the hard-line government in Hungary, helped elect a center-right government in Poland. Xenophobic backlash and sometimes violence against refugees have occurred in Germany, Sweden, the UK and other countries.

The rush of refugees raises fundamental questions of European identity. In one way, they are a boon for a continent whose low birth rate and aging populations needs workers. But Europe has never been in modern times an immigration-welcoming society, unlike the United States which has welcomed wave after wave of immigrants and, if not perfectly, integrated them into US society. European society sees itself as largely Christian, and worries over the influx of Muslim refugees. Problems of integration of Muslim immigrants in the past, eg: in France and the UK, have been sharpened by terrorist recruitment among this population, aggravating the current concerns.

The refugee problem has thus returned to Europe, and it cannot any longer be kept far away as a purely humanitarian enterprise. As Europe struggles to address it, it is turning to some of the trouble spots to see if it can stem the flow. Until recently, Eritrea accounted for the second largest number of refugees reaching Europe. The EU was, in late 2015, finalizing a five-year €200-million development for that country, setting aside for the moment previous objections to Eritrea's human rights record. But countries at war, like Syria and Somalia, do not present such opportunities. Europe thus faces fundamental issues of identity, political change, and economic challenges from this change in the international scene.

Conclusions: Refugees are both the seeds and symptoms of conflict. They can be treated in the short term as a humanitarian problem. But that is less and less sufficient as their presence grows and the capacity and willingness of other countries to receive them diminishes.

Oddly, while neighboring countries in the Red Sea and Horn have the greatest immediate burden of refugees, with all the complications they present, this alone has not led them to address more effectively the conflicts that have produced them. As a result, other interests seem to prevail.

For example, the civil war in South Sudan had produced more than 600,000 refugees by the end of October 2015, flowing into nearly all the neighboring countries. But competition between two of the Intergovernmental Association for Development's (IGAD's) principal members — Sudan and Uganda — led them initially into essentially a proxy war in South Sudan. And although Ugandan forces have now withdrawn, the two continue to compete for influence over the country through their support for one or the other contending parties. These and other differences between IGAD members have kept them from agreeing on an arms embargo or other harsh sanctions on the combatants. As long as such disunity continues, IGAD's ability to enforce the peace plan upon the contending parties will be difficult.

On a larger scale, the African Union is struggling to take ownership of its peacemaking responsibilities. However, the AU remains almost totally dependent upon donors for not only its peacekeeping missions but its regular budget as well. Recently it committed to provide 25 percent of the peacekeeping costs by 2020, for those enforcement or counter-terrorism missions that are not appropriate for the UN — such as AMISOM — seeking UN assessments for the balance. But how it will raise even its 25 percent remains uncertain. With oil prices half of what they were in 2014, major African providers such as Algeria, Nigeria, and Egypt are balking at picking up more of the costs.

By separating out the humanitarian costs and assigning them to the international community; by paying insufficient attention to the longer term political, economic and environmental costs of refugees and IDPs; and putting such matters as peacekeeping into a separate category of concern, the full costs of conflicts are not evident and do not sufficiently force strong action to end them.

Until these are placed more squarely in front of all those affected, and especially the foregone opportunities which such costs represent, the full force of political pressures will not be brought to bear on these conflicts.

McDermot
Will & Eme

VIII

Transforming RedMed Infrastructure and Trading Patterns

The Potential for the Red Sea Region to be a New Center of Global Growth

Andrew Pickford

THROUGH THE 2010S, INTERNATIONAL attention on the wider Middle-East region has been centered on Iraq and increasingly Iran. Unnoticed by many, there have been a number of developments which could shift the center of economic activity westwards towards the Red Sea.

A number of interlinking economic and energy trends has meant that traditional trade routes, some millennia old, are now re-emerging. This chapter examines the dynamics of the Eastern Mediterranean and specifically the energy policies of Israel and Egypt. Both of these countries have experienced radical changes to their energy frameworks and could benefit from exploiting major new, offshore gas discoveries.

While Israel and Egypt may become surplus producers of natural gas, many Persian Gulf nations are likely to be more dependent on natural gas imports. Saudi Arabia's challenges are particularly acute as its economy is buffeted from lower oil prices and the distortion caused from subsidies, which together are causing major fiscal strains. Should Egypt and Israel be able to maintain an attractive investment environment and potentially share infrastructure; economic growth will be further stimulated. This could transform the entry point of the Suez Canal and potentially make it a new global business hub.

Economic Drift, Dislocation, and a New Economic Landscape

In discussions over energy, there is a tendency to skip over fundamentals. While the link between GDP growth and energy intensity has been weakening, there still is an important correlation. This is relevant to Persian Gulf nations which derive a large amount of their national income from oil sales. Economic growth and energy prices will also influence investment and development of natural gas fields in the Eastern Mediterranean.

At a macro-level, the International Monetary Fund predicted global growth to be 3.3 percent in 2015. Despite optimism for 2016, global growth could slow down even further. At the country and region-level, there remains a range of challenges which have been partially masked by China's entry into the global economy. However, China is now much different to the 2000s when it was an engine of growth. Its greater focus on internal consumption means that the high rates of export-led growth will likely not be repeated. In Japan, the "Abenomics"[27] experiment is still underway. Despite demographic and geopolitical trends, it is possible that there could be a transformation of the Japanese economy. The European Union faces a number of fundamental issues, with the lack of control over migrant flows a symptom, rather than a cause, of deeper structural prob-

27 Abenomics refers to the economic policies promoted by Japanese Prime Minister Shinzô Abe and is based upon the "three arrows" of fiscal stimulus, monetary easing and structural reforms.

lems. It is difficult to see how the internal rifts and embrace of environmental policies will produce anything more than mild growth prospects at best. This may mean the common currency and treaty arrangements could disintegrate. Another important grouping, the Association of South-East Asian Nations (ASEAN), has the potential to coalesce into a key economic zone, evidenced by ambitious plans to exploit low cost energy and leverage its natural resources.

Through the 2020s, parts of Africa may start to emerge, although this will probably be uneven as internal migration to larger cities increases their importance in internal political processes and results in imbalances favoring consumption before a stable economic base can be established. While the US economy is in reasonable condition, its economic recovery has not been producing as many jobs as expected and low cost shale oil and gas has masked the ossifying influence of a more government dominated economy. Furthermore, efficiency gains in transportation vehicles, technical advancements and changing consumption patterns means that US oil demand growth has been very slow. Talk of a "peak supply" challenge has shifted towards a "peak oil-demand" paradigm.

As individual nations and regions adjust to new trading modalities and energy patterns, the global economy is also evolving. It is currently undergoing a digital transformation, which has the potential to unlock productivity gains to a greater degree than the Information Technology era of the 1990s. This is creating efficiencies, and likely to fundamentally change employment patterns in developed nations especially as automation accelerates. For countries in the Red Sea region, this means the transition of the Asian "Tigers" from poor, undeveloped nations to developed affluent countries will not necessarily be repeated. The experience of Singapore, Hong Kong and South Korea may no longer be suitable reference points because automation and the concentration of manufacturing in China have reduced options for low-cost economies to follow traditional development models.

Growth in Red Sea nations will need to be based on 21st Century realities. This may involve a more important rôle for agricultural production and other secondary and tertiary industries. It could also be driven if the region regains its position as a key conduit of global trade flows. Given the youth bulge of these countries; prosperity, or lack thereof, will influence internal stability. Leaders across the Maghreb and Middle East are carefully weighing up the need to undertake economic reform but also the need to maintain popular support following a series of power changes known in the West as the "Arab Spring".

Navigating the balance between maintaining popular support and necessary changes to state budgets, subsidies and services could produce uprisings, protests and internal fragmentation. Managing expectations will be essential in this phase. Successful governments will be able to reinvigorate national pride, iconic symbols and a founding narrative. Channeling these efforts to productive endeavors is critical, but made easier by the civilizational history of nations within the Red Sea region. If these sentiments are unleashed as chauvinistic nationalism, it could increase tensions and potentially lead to conflict.

Evolving Trading Patterns

Throughout the past thousand years, there has been competitive tension between land and sea routes, with air-based transport a very recent phenomenon. The geography of the Red Sea region has, at times, influenced global trends. Understanding the interplay between European, Middle-East, Asian and African empires, as well as local powers is helpful for appreciating these drivers of logistics and transport patterns. The Red Sea region was on the edges of Silk Road trading routes and Roman Empire road networks. Centuries after the last treasure ship was recalled to Ming China and the last legionnaire returned to his home province, the physical and cultural legacy of these networks are now re-emerging.

Historically, land based networks of Eurasia were the main routes of global trade. The rise of the Ottoman Empire and improved naval technologies in Western Europe meant that, in a short period, sea-based transport was competitive. By the 1700s, it was the dominant conduit of global trade. In 2016, sea-based trade remains the main way to transport commercial goods. However, a pan-Eurasian transport network may shift some of the freight task (in percentage terms) back to land. A fully integrated Eurasian road and rail network could reduce transport times from Guangdong to Portugal which would create price competition with sea routes.

The Red Sea region increased in importance for global trade when the Suez Canal was first opened in 1869. European-Asian trade and the rise of Middle-East oil helped drive the region to a more critical rôle, most often identified during crises. A shift back to land-based routes would not necessarily diminish the rôle of the Suez Canal zone. Its rôle as a hub could even be enhanced if it captures more of the North-South trade from expanding African and Central Asian economies.

China's One Road, One Belt strategy is partially re-creating old Silk Road and Indian Ocean monsoon trading patterns. To the west of the Red Sea, the Roman road networks remain the template for a number of key transport routes. As Eastern Europe countries grew after the end of the Soviet Union, the interconnected pan-European trade systems re-emerged.

Aside from the physical routes of goods, there is the potential for geo-economic *blocs* to form and influence trading patterns. The US-led Trans Pacific Partnership and PRC-led Shanghai Cooperation Organization may be the beginning of these new *blocs*. The membership and establishment of the China-led Asian Infrastructure Investment Bank is further evidence of the declining influence of the post-World War II financial and economic structures. If *blocs* form, Red Sea countries will need to astutely manage their position so as not to be excluded from commercial opportunities.

Energy Trends and Low Prices

There is a strong connection between energy prices and the trajectory of economies within the Maghreb and Middle-East. This varies from country to country and not only influences budgets, but also industrial policy. In Saudi Arabia the petroleum sector accounts for roughly 80 percent of budget revenues and approximately 45 percent of GDP. Lower prices reduce spending options for governments which then must draw down reserves or borrow. For nations without energy reserves, the cost of importing energy for transport and electricity can weigh heavily on stretched budgets.

Over the past decade there was a demand shock which pushed oil prices close to $150 a barrel which was followed by a period of oversupply. The cumulative impact of the US shale revolution, OPEC strategy and China's slowdown has resulted in a period of lower oil prices. This may extend to 2020. The changing nature of shale oil drilling and "banking" oil is likely to create a new oil price ceiling. A Manhattan Institute research paper found that: "continued [shale-related] technological progress, particularly in big-data analytics" would "ultimately yield break-even costs of $5-

$20 a barrel, which would put costs on par with those in Saudi Arabia."[28] In this scenario, onshore US shale operations become the global, oil swing producer. This rôle would be governed by market forces rather than cartel-type arrangements.

In the broader energy sector, it is expected that the trend towards natural gas as an electricity source would continue for a period of time. The flexibility of natural gas within the electricity system makes it attractive to developing nations which are dealing with urban pollution and facing large increases in energy demand. In Western countries, international and domestic environmental legislation and regulation will likely favor an expansion of natural gas at the expense of coal.

Liquefied Natural Gas (LNG) — natural gas which is cooled to liquid form and transported over oceans on purpose built ships — is expected to grow in importance over the next two decades. However, LNG in East Asia will remain in surplus for some time. This is influencing downward price pressure in the three key global reference points and hubs: "Henry Hub" (US), National Balancing Point (United Kingdom), and Platts JKM™ (Japan Korea Marker). There is a convergence of global gas prices, but not yet a fluid global market. For future natural gas producers, the depressed price for natural gas means that investors are for the short- to medium-term very cautious about commissioning new projects.

Israel as an Energy Island

Within the Eastern Mediterranean, the position of Israel is relatively unique due to geopolitical factors which limit options for it to create energy security and achieve econo-

28 Mills, Mark P.: *SHALE 2.0: Technology and the Coming Big-Data Revolution in America's Shale Oil Field*, No. 16 May 2015, Center for Energy Policy and the Environment, Manhattan Institute.

mies of scale within electricity systems by integrating with neighboring countries. As a result, the priority for Israel is achieving self-sufficiency and security of supply. For this reason, despite its land borders to other nations, it is sometimes referred to as an "energy island".

In the past, Israel relied on imported oil and coal for electricity generation. However, the past decade has seen a transition to natural gas. This has followed the worldwide trend of increasing natural gas use in electricity generation. In Israel, the transition was advanced rapidly because of discoveries in the 1990s which were viewed by decision-makers as a solution to security concerns. Given that the state was heavily involved in the energy sector, predominantly through the Israel Electric Corporation, the transition could occur at a fast rate. However, it was not until 2005 that natural gas entered the energy mix when production began at the Yam Tethys offshore project.

The shift to natural gas was rapid and the trend is likely to continue for the next few decades. In 2001 coal was used to generate about 78 percent of Israel's electricity. By 2010 the share of natural gas in the electricity mix reached 36 percent with coal at 61 percent. By the first quarter of 2015, State-owned Israel Electric Corporation —which accounts for 81 percent of the electricity production in the domestic market —reported that natural gas comprised 42.2 percent of electricity generation with coal decreasing to 57.3 percent. Israeli gas demand is expected to grow over the next two decades, primarily driven by electricity generation. Official forecasts by the Israeli Ministry of Energy and Water Resources envisage natural gas to account for 60 percent of electricity generation in 2027 and 68 percent in 2040.[29]

The re-weighting towards natural gas shifted national

29 Israeli Government, Min. of National Infrastructure, Energy and Water Resources, *The Natural Gas Sector in Israel*, http://energy.gov.il/English/Subjects/Natural%20Gas/Pages/GxmsMniNGEconomy.aspx [Acc. November 11, 2015.]

priorities. Aside from building a new domestic natural gas industry, there was also the need to secure supplies from regional countries. A 2005 import agreement with Egypt accelerated the growth of Israel's gas industry. Gas was imported through a pipeline operated by the jointly owned East Mediterranean Gas Company. By 2010, Egypt was supplying Israel with 43 percent of its natural gas needs. Following internal political unrest in Egypt, the pipeline was sabotaged and gas flow became sporadic. In March 2012, the Egyptian Government cancelled the agreement, bringing imports to a halt. As 40 percent of Israel's power generation capacity relied upon natural gas, the lack of feedstock caused major challenges to the grid operation. With peak demand approaching the installed 12,370 megawatts of capacity, the result was sporadic brownouts and power quality problems across the country.

In 2013, Israel entered into a short-term contract for two LNG loads a month. These LNG imports served as a stop-gap measure while domestic production ramped up. This short term response was similar to Japan when it took nuclear reactors offline after the Fukushima Daiichi nuclear disaster. LNG served as a stop gap measure while there was a review of energy policies.

Israel's Offshore Gas Fields

Over the past decade there have been a number of major offshore gas discoveries in the Eastern Mediterranean within Israeli waters. Israeli political leaders saw this fuel source as a long term secure supply for Israel, as well as an export revenue earner. Like many countries which discover a large natural resource under their control, popular demands and calls for rent-seeking[30] make exploitation diffi-

30 Rent-seeking: When a company, organization, or individual uses its resources to obtain an economic gain from others without reciprocating any benefits back to society through wealth creation. — Investopedia.com.

cult. This is especially the case for countries without a long term history of resource extraction. While touted as a solution to a range of internal challenges and security concerns, these larger fields were not simple to exploit and were subject to forces outside the control of the Israeli Government.

In 2009, a 10.8-trillion cubic feet natural gas field, Tamar, was discovered offshore. Leviathan; a field almost twice the size, was found in 2010. At present, Tamar field supplies the bulk of Israel's natural gas needs, including 40 percent of electric power. The gas is funneled through two pipelines to the Tamar platform. It is then piped to a terminal feeding several power stations which convert the gas into electricity to power Israel's electric grid.

As has been the experience with large offshore gas fields, these deep water projects are capital intensive and require advanced technology. This requires a stable and predictable regulatory regime. Given the volumes of gas from Leviathan are well in excess of the domestic requirements of Israel, there is a need to consider export options. Israel cannot export gas-generated power simply by transmitting the electricity through its neighbors' power grids. One alternative is through pipelines or via LNG.

Long and expensive pipelines to Turkey or Greece are unlikely. However, the resumption of Israeli-Turkish relations and increasing tension between Turkey and Russia (a major supplier of natural gas to Turkey) may make the sub-sea pipeline slightly more attractive.

While Jordan is committed to take Leviathan gas over a 15-year period, this may not be enough to underwrite project development. The other option to underwrite a major new project is to sign long-term LNG supply contracts. However, LNG export prospects at the moment are limited. A large increase in supply has depressed prices, with oversupply expected through to the early- and even mid-2020s.

Floating LNG plants, similar to those developed in Australia for Shell's Prelude project, present one option. However, these are expensive and the technology is still relatively new. The so-called window for new LNG projects has largely closed for the next few years. Investments in this area tend to be lumpy and mega-projects with large budgets which are heavily scrutinized against alternative options across the world.

After the discovery of the Leviathan field there was an internal Israeli debate about the optimal development path to generate funds for the state while also preserving supplies for future generations. For international oil companies, these extended deliberations heightened the sovereign risk calculations which saw the Australian energy firm Woodside withdraw from the project. There is not yet an overall consensus on the development and export strategies. Outside of a crisis, there does not seem to be an outcome where the development of Leviathan (and other undiscovered mega-fields) will occur in the short-term.

Israeli debates over natural gas development and surpluses from mega-projects is not unique. Nonetheless, the evolving taxation, anti-trust and regulatory environment has reduced investor confidence. This is similar to the Canadian province of British Columbia, when supposed windfall profits were calculated in advance of projects being approved and there was a period where the level of the tax rate was not clear. Just as the LNG window for Canada has largely passed, the options to develop Israeli fields is more difficult. This is partly caused by the transformation of Egypt's investment landscape; newly discovered offshore Zohr gas mega-field; and low gas prices.

Energy analysts expect that the Egyptian discovery of Zohr will reduce the likelihood that Israel will supply Egypt with gas. Production from Egypt's new field could start in

2017, some two years earlier than when Leviathan is targeted to go on line. Israel needs substantial private investment and external markets to make the Leviathan and future gas projects a reality. Even if an agreement on Leviathan can be reached, further delays to the project are likely. Under these new conditions, Israel will be dependent on developments within Egypt.

Egypt's Challenges and Opportunities

The strategic position of Egypt in the Red Sea is well known, especially due to the pivotal rôle of the Suez Canal. The Suez Canal earns the Government about $5.3-billion in annual revenues. An expansion of the canal could increase annual revenue to $13.2-billion by 2023. There is also a plan to build the $55-billion Suez Economic Development Zone by 2030. This development zone may become the beginning of a larger industrial complex and in time become economically more important than the Suez Canal.

In 2030, based on projections, the population of Egypt is expected reach 116-million, an increase of 42 percent from 2012. Egypt has a growing, youthful population, low labor costs and is close to affluent Western European markets. This would make it an attractive place to build a manufacturing industry. Investor confidence can be sustained and pro-market and pro-growth policies can become the norm and embedded. A continuing challenge for the Egyptian government is to repair fiscal settings and encourage investment.

Energy subsidies have been costing the Egyptian Government approximately $18.8-billion or about 10 percent of the GDP. If electricity was included with fuel subsidies, this equated to around 30 percent of the budget in 2013. Reforms announced on July 4, 2014, were designed to reduce energy subsidy spending to $6.2-billion in the

2014-15 fiscal year.[31] This is more than the Suez Canal earns each year. Lower energy prices provide a valuable window to deal with subsidies. However, too fast a change can produce backlashes either at the ballot box or on the streets.

Egyptian Offshore Gas Fields

In 2012 around 75 percent of Egypt's electricity was produced by using natural gas. Market distortion caused by the price paid by Egypt to producers, and the official tariffs price for sales has impacted natural gas production and consumption rates. Over the past five years, Egypt's annual energy production grew on average by one percent compared to annual average consumption growth of 5.3 percent, increasing the gap between supply and demand. During the first part of the 2010s, there was a decrease in the domestic supply of gas. Delayed payments to upstream companies coupled with the low prices offered for supply reduced investment and output. The electricity grid struggled to keep up with demand in a difficult operating context.

Reforming energy markets in any country is extremely difficult, with heating, cooling and cooking providing a direct connection between abstract economic matters and physical comfort. Just as ensuring a ready supply of 'bread and circuses' was a major priority of Roman administrations, modern governments in hot climates need to be cognizant of bread and air-conditioning. If access is rationed by price or other mechanisms which are perceived as unfair, restless populations will react.

When Abdel Fattah el-Sisi took control of the Egyptian Presidency, the economy and country was in a perilous state. Should Pres. Mohammed Morsi's Administration have remained in power, economic (and state) collapse

31 James, Laura M.: *Recent Developments in Egypt's Fuel Subsidy Reform Process*, International Institute for Sustainable Development, April 2015.

would have been a real possibility. Aside from maintaining order and keeping the organs of state functioning, Pres. el-Sisi understood that economic growth and opportunity was necessary to maintain the unity of the Egyptian state. He has aimed to achieve a recovery through: further support from the Gulf States; energy subsidy reforms; and increased foreign investment.

Pres. el-Sisi's Government set a medium-term macroeconomic policy framework, covering the fiscal years from 2014/15 to 2018/19. This aims to reduce the fiscal deficit from 12.6 percent to 8.5 percent of GDP, and public debt from 97 percent to within 80 to 85 percent of GDP, in five years. The framework states that "reform of energy subsidies will account for the lion's share of the planned fiscal adjustment, with subsidies on all fuel products (other than LPG for needy households) and electricity to be phased out over the medium-term."

The availability of natural gas and a functional power system were important priorities for the el-Sisi Government. Stop gap measures included importing LNG. In November 2014, Egypt leased a floating LNG storage and regasification terminal for five years. The Egyptian Natural Gas Holding has signed contracts for contracted 90 cargoes of LNG to supply the vessel in the coming year from a range of trading companies and gas producers including; Trafigura, Vitol, Noble, and Sonatrach. In March 2015, Gazprom signed a contract to deliver additional cargoes until 2020. Additional contracts will probably be secured in 2016.

Like Israel, Egypt's energy and economic future will be tied to securing long-term supplies of competitively priced gas which is located offshore.

In the first decade of the century, when Egypt was relatively stable and prior to recent unrest, there was considerable offshore exploration. During this time the country's gas reserves more than doubled.

With proven reserves of 1.8-trillion cubic meters, Egypt has 40 percent as much gas as Algeria. However, while production increased in the early 2000s, according to BP, output peaked in 2009. By 2013 production had declined by nearly 10 percent. The loss of production was caused by a range of factors, not least the low price paid by the Government for the final product. This was far too low a price to allow discoveries in the deep-water Nile Delta area to be developed.

To increase energy supply, the el-Sisi Government accelerated the payment of overdue fees to international oil companies to encourage them to increase their investments in development and production activities.

Egypt has signed 53 new exploration agreements with a total estimated value of $2.9-billion. Improved terms have seen BP proceed with its West Nile Delta development which was scheduled to begin in March 2016: a $12-billion investment. By itself this will increase the country's gas production by 25 percent. Italian firm Eni has also signed a framework agreement covering $5-billion of investment in oil and gas developments over four years.

A great deal of attention followed the discovery of the Egyptian gas mega-field Zohr by Eni. Zohr is reportedly around 40 percent larger than Israel's Leviathan field. Exploration work will continue through 2016 which will provide a better indication of the full size of the field.

To date, Zohr is the largest find in the Mediterranean. It lies far below the seabed in waters almost 1,500 meters deep, but is within reach of existing infrastructure, although still expensive to develop.

Zohr's estimated gas would add almost 50 percent to Egypt's existing reserves and would be enough to meet

Egyptian demand at current levels for 17 years.[32]

In mid-October 2015, Pres. el-Sisi and Eni CEO Claudio Descalzi confirmed a commitment to accelerate the production start-up of the Zohr field. This could potentially occur in 2017, but would require a range of difficult hurdles to be overcome. Eni's CEO and Sisi further discussed the possibility of setting up a gas hub in the eastern Mediterranean Sea aimed at making the Egyptian transport and export facilities also available for other nearby discoveries.

Language referring to Zohr as a "supergiant" and a game changer, is not helpful. Much of the attention has focused on the impact on the viability of Israel's Leviathan. However, with one or both coming online in the next few years, the major beneficiaries could be Southern European gas buyers. Rising expectations from the public may also force sub-optimal policy decisions.

Egyptian energy reform has generally shifted towards a more market-based and pro-investment approach. This will help address increasing demand for electricity and broader decisions about industrial development. However, policies on renewable energy will create future budget burdens. The 2014 announcement of a feed-in-tariff and 4,300 megawatt quota for installed solar and wind was regarded positively in the West, but could turn out to have a negative impact on the electricity system and represent a new cost to government.

Gas Shortages in Persian Gulf Countries

Most Persian Gulf countries are having trouble meeting growing gas demand at home while sustaining exports. In the period of high oil prices, with growing economies and populations; subsidized domestic gas and policies of energy-intensive industrialization were financially viable.

32 Gavin, James: "Israel struggles with timing", *Petroleum Economist*, September 21, 2015.

This was marked by the unusual circumstances of Kuwait importing a load of LNG from Western Australia.

The drop in oil prices has meant that discounted gas and economically inefficient uses of energy are draining state budgets. However, with a coalition of subsidized industries and consumers, there are barriers to meaningful adjustments of prices to market levels. Accordingly, the region's state-owned companies, with assistance from international companies in some places, have raised gas output significantly in Saudi Arabia, Abu Dhabi, Iran, and Oman. Nevertheless, it is not keeping up with demand. Given the transitions and unrest in the region, there has been nervousness about fundamental reform of prices. The heavily subsidized and low price of gas in Saudi Arabia means that there is a fundamental mismatch between supply and demand. All of Saudi Arabia's gas output is consumed domestically and it still has to run half its power sector on oil.

There is little gas being exported from the region, with exports expected to decline. Iran and Qatar are the only countries where exports could increase. Neither is expected to make radical changes to their existing polices, which have been delayed since the drop of gas prices. Other countries such as United Arab Emirates and Kuwait are expected to increase their LNG import capacity; the regional net effect may continue the declining trend. Prospects of North African exports vary. Algeria's declining production and internal challenges are not expected to alter this trajectory in the short- to medium-term.

In analysis over the future of the region, the continued existence of Saudi Arabia in its current state is not guaranteed. Prolonged low prices of oil will erode the financial reserves of the Kingdom. If an article in the peer-reviewed *Journal of Petroleum Science and Engineering* is correct, Saudi Arabia could potentially experience a peak in oil production in 2028.[33] The peak would be followed by inexorable decline. Such a scenario means that even if oil prices moderately recovered in the 2020s, the window for a transformation of Saudi Arabia's economy would be very short. At best, eight years of strong prices would be needed to help prepare for a scenario where oil was a much smaller part of the economy. It is possible that the social compact could dissolve, and the State would either break up or shrink after neighbors reclaim areas of their traditional influence. Depending on Iran's strategy, as Saudi Arabia weakens, Tehran could foster unrest in the Kingdom's eastern province.

In addition to the stability of Gulf nations, the energy policy of nations across the broader Middle-East and North African regions is important to Red Sea countries. The 2020s may result in a new dynamic as Iran re-asserts itself and the People's Republic of China becomes more enmeshed in internal dynamics of regime survival. This would probably not follow the deliberate, but relatively peaceful displacement of UK strategic interests by the US. Even if the core of the Middle-East adjusts back to pre-colonial areas of influence rather than back to the framework of modern nation states, the Red Sea region could become a place of relative peace and stability with economic growth based on more balanced economies.

As Iran becomes more integrated with the world economy, the influence of the clerics will likely decline as business interests and other expanding groups have more influence on national policy. Despite current Iranian-Israeli and Iranian-Egyptian relations being at a low ebb, a reversal to historical patterns of interaction is possible. Iran and Israel have had millennia of positive interaction and the last Shah of Iran was an associate of Egyptian President Anwar Sadat,

33 Mohsen Ebrahimi and Nahid Cheshme Ghasabani: "Forecasting OPEC crude oil production using a variant Multicyclic Hubbert Model", *Journal of Petroleum Science and Engineering*, April 20, 2015.

and his first wife was Egyptian. A reinvigorated Iran and economically dynamic Egypt and Israel could start finding common areas of strategic interest during the 2020s. Under these conditions, the current Western focus on the nuclear weapon program of Iran will be less relevant. As the PRC builds its influence and economic presence in the Red Sea and Persian Gulf, Washington and London will have limited ability to drive the international agenda. A more pressing issue may be under a scenario where India starts to assert itself and the Persian Gulf and hinterland once again become enmeshed in a great game between emerging global powers.

Prospects and Analysis

Despite the large gas fields found in the Eastern Mediterranean, it is not inevitable that these reserves will be developed. It will depend on a range of internal and external drivers and the fact that these deep, offshore fields require a large mobilization of technology, capital and engineering expertise, often beyond the capability of a single state. Given exploration efforts and the investment attractiveness of offshore Egypt, there will probably be further discoveries which add to the resource base. Before international oil companies become involved they need the right gas prices, fiscal terms and incentives to operate efficiently and access the best technology. Floating LNG plants have given producers some more flexibility in terms of sovereign risk concerns as they are not fixed to a particular nation and, in extreme circumstances, can be moved. Developing large offshore natural gas fields is challenging and the large investments mean that investor confidence needs to be built with a number of guarantees, legislative mechanisms and support required before projects are sanctioned. This will be particularly the case up to at least 2020.

The interplay between gas investments and decisions in

Egypt and Israel should not be understated. This relates to markets, access to facilities and pipelines and also export routes (Suez). Greater offshore assets will need further protection, surveillance and are already pushing the use of aerial and underwater drones. Like Australia, this changes security policies and defined areas of national interest. As larger, more expensive offshore assets are built, the logic of economic and security cooperation between Israel and Egypt will increase. Given low energy prices, internal Israeli debates over resources, and Egypt's recent history of power transitions, cooperation will probably accelerate in the 2020s or when the price of gas recovers.

The Egyptian energy and economic reforms are promising, but there is still a lot to do to balance the popular demand for "bread and circuses". Despite positive advances on fuel and gas subsidies, renewable subsidies will probably create a long-term liability for the budget.

Even with a reorganization of the power structure in the Middle-East, Egypt and Israel could start to build an understanding around a more durable Eastern Mediterranean economic zone. Both would benefit from an economic and gas trading hub in Egypt. Due to Israel's experience and instinct for energy security on its terms, this will not be their first preference. However, a stronger and internationally focused Egyptian economy will be more predictable than one which is approaching disaster. If countries in the Red Sea region start to pursue their own interest, and not be proxies for external power competition, there is potential for it to become a new center of global growth.

IX

The Smaller RedMed Economies

Somalia, Somaliland, Djibouti, Eritrea, Yemen, and Jordan

Gregory R. Copley

THE SIX SMALLER RedMed littoral economies — Jordan, Yemen, Somalia, Somaliland, Djibouti, and Eritrea — stand apart, for economic reasons, from the major regional powers, but each has the potential to exert disproportionate strategic leverage over the security and stability of the region because of their respective geographic locations.

Indeed, it could be argued that all coastal access to the Red Sea/Suez sea-lane is potentially vital, despite the fact that these six smaller economies represent a total of only some $85-billion in combined GDPs out of an overall $1.5-trillion in GDPs for the overall RedMed theater.[34]

This disproportionate geopolitical influence, when compared with economic size, is now particularly evident as Djibouti turns from a backwater to a major center of strate-

34 GDPs (2014 except where noted): Jordan $35.827-billion; Yemen $35.95-billion (2013 est.); Somalia undetermined, but possibly estimated at $5.698-billion, according to the IMF ($917-million in 1990); Somaliland $1.4-billion (2012 est.); Djibouti $1.582-billion; Eritrea $3.858-billion. Approximate total GDP in 2014 estimated at around $85-billion, out of the RedMed zone's estimated $1.5-trillion.

gic competition, and as Yemen battles to determine whether it remains a unified state.

Like Yemen, the four smallest economies on the African littoral of the Red Sea — Somalia, Somaliland, Djibouti, and Eritrea — offer some of the most significant challenges and opportunities for the security of the Bab el-Mandeb Strait.

Somalia and Somaliland at present represent a power vacuum along the shores. And geopolitics, like nature, abhors a vacuum.

We saw, however, that under certain conditions, the vacuum is replaced by a problem. For example, the surges in the conflict in Somalia spill into Kenya, or the fact that a lack of governance in Somalia allowed piracy to flourish in Puntland until it was arrested by a concerted international response. Under such conditions, the calm of the power vacuum is replaced by a challenge which threatens the global sea route.

When that occurred in the form of the Puntland-based piracy, we not only saw the significant military response by the United Nations and the African Union to address the Somalian civil war, we also saw the creation of the 30-nation Combined Maritime Force (CMF) and the establishment of Combined Task Force 152, which brought, for example, a permanent naval presence to the north-western Indian Ocean. Other non-CMF players, such as Japan and the People's Republic of China took up station at the mouth of the Red Sea.

It is obvious to see the logic in the British decision to control both sides of the Bab el-Mandeb, with the Aden Protectorate being joined under the Union flag in 1886 with the creation of the British Somaliland territory.

Here is a broad and brief strategic view of the impact of these six RedMed states.

Let us begin with Somalia and Somaliland.

Somalia and Somaliland

Let me preface my remarks by saying that, although the two Somali states have some linked historical and cultural aspects, their geography, and the way in which their clans evolved has made them different geopolitical entities. Attempts by the international community to force a reunion of the two would likely create a painful outcome as we saw after 1960.

Arguably, the brief iteration of a so-called "unified Somalia" was very much the product of World War II, which had left Britain in control of the formerly Italian-controlled region as well as British Somaliland. The British fostered the concept of a unified Somalia, even though historical social and political developments had determined different identities.

Considerable strides have been made in stabilizing some parts of Somalia (that is, what was once Italian-occupied Somalia), but there is little evidence of a lasting solution which would see a productive, stable state. There is even less opportunity for that lasting solution if the international and regional communities fail to recognize the fact that Somalia is not as it was in 1960, when the two Somalilands joined to create Somalia. That experiment failed as profoundly as the brief creation of the United Arab Republic which linked Egypt with Syria from 1958 until 1961.

Moreover, if we recognize that the fundamental basis of clan organization in Somalia — that is, the former Italian Somaliland — has been around water wells, then we can see that, in a country thus far bereft of other resources, water is the great challenge. And yet there is no great initiative to transform the water challenge facing the Somali peoples.

Indeed, water is the great challenge for virtually all of the RedMed theater. But in the case of Somalia, it is unlikely that clan warfare can be eradicated — and the benefits of agriculture introduced — without a strategic approach to introduce water. Every other attempt to stabilize Somalia is a band-aid solution: short-term, reactive in nature. This gets to Gen. Aliyu Mohammed's question as to whether the recent history of UN and international community peacekeeping is resolving problems, or postponing them for a later generation.

Not that there is an alternative to the forcible suppression of lawlessness when it impacts a vital world resource such as the Suez-Red Sea SLOC.

The multinational combined maritime task force, CTF-152, and other initiatives have helped calm the piracy from Puntland, leaving that area of Somalia once again as an impoverished backwater. Much of the revenue which Puntland-based pirates won from their activities, however, was funneled into investment in Kenya, exacerbating the linkage of criminal activities from Somalia to Kenya.

African Union projection of long-term peacekeeping into large areas of Somalia had created a sense of calm around Mogadishu by mid-2015. But the reality was that the problems of Somalia had not been solved, and there was no real prospect in sight for a stable Somalia state.

The Republic of Somaliland, which moved from a period of considerable progress after withdrawing from the union with the former Italian Somalia in 1991, has very significant strategic assets on the Red Sea, but is now in the hands of a dysfunctional government. The irony was compounded by the fact that many of the factors which contributed to the refusal of the regional bodies to recognize the sovereignty of Somaliland had, by 2015, evaporated.

But that is happening just as Somaliland seems to be descending into chaos. The situation began to unravel rapidly after Pres. Silanyo's disastrous visit to Ireland and London in September 2015, ostensibly *en route* to the United Nations General Assembly on September 28. He did not go on to New York, but remained for several weeks in London for medical treatment and spoke with the expatriate Somaliland community, who were hostile to him. It was clear on that trip that his dementia had become pronounced. He could neither hold his papers, nor speak rationally. So when he returned to Hargeisa in late October, it became clear that he could no longer even pretend to lead the country.

His wife, Amina Sheikh Mohamed Jirde, long the power behind the scenes, with two of her lady companions, ensured that the Deputy Chairman of the *Hisbiga Kulmiye* (Solidarity Party), Moussa Bixi (Bihi) Abdi, would be named as Silanyo's successor. This angered the key ministers who had been essentially running the Government, and seven of them resigned on October 26, including Minister of Presidency Hersi Ali Haji Hassan, who had been close to some of the *jihadi* movements; Minister of Foreign Affairs Mohamed Abdullahi Omar; the Minister of Labor; the Minister of Justice; and the Minister of Public Works.[35]

It remained a moot point whether new elections could even be held, given the collapse of the Government, which had, in any event, delayed elections beyond their scheduled date in July 2015 because voting rolls allegedly could not be

35 The former Minister of Presidency in Somaliland, Hersi Ali Haji Hassan (Somali: Xirsi Xaaji Xasan), who resigned on October 26, 2015, along with six other Cabinet members, began moving rapidly to transform the country into a pan-Somali Islamic state. Sources close to the former Minister, who was part of Yemen's *al Islah* (Congregation for Reform) movement, was moving to raise large amounts of money to finance a movement which would transform the governance of Somaliland into something like the Islamic Courts movement which prevailed briefly in Somalia, or the Islamist movement of National Islamic Front (NIF) leader Hassan al-Turabi in Sudan. Hersi Ali Haji Hassan, working with former Finance Minister Abdel Aziz Somale, were banking on the system of *tazkia* (literally purification) through which funding was made available to religiously sympathetic supporters to fund businesses. As a result, recipients were being pressured to contribute funds to the new movement. The trend, then, seemed to be for Hersi Ali Haji Hassan to be working to finance what would be an armed movement, which would draw on radical Yemeni as well as Somalian resources.

compiled. It seemed likely that Moussa Bixi Abdi could be "installed" as President to replace Silanyo, without an election, but with the rump Parliament merely rubber-stamping the position.

The origins of the problem, however, go back to Somaliland's original reassertion of its independence in 1991. Somaliland's renewed independence from the Union of Somalia was seen by Egypt and the Arab League as potentially benefitting Israel and hurting Egypt, but that concern is now non-existent. Indeed, Egypt's growing ability and need to project maritime power and control over the Mediterranean and Indian Ocean access points to the Suez Canal could make the Somaliland port of Berbera an attractive basing option for the Egyptian Navy, particularly with the proposal to station at least one of its two new *Mistral*-class helicopter carriers in the Red Sea.

Taking the strategic view, it is clear that the delay in international acceptance of Somaliland's withdrawal from the Union of Somalia has meant that Somaliland has finally now fallen into the hands of those who see its isolation as a convenient safe-haven for criminality. It is clear why Somalia would wish to assert claims over Somaliland: it is the most significant agricultural and resource-rich area of the two-state area. It is equally clear why Djibouti would remain cool to a fully sovereign Somaliland, with which it shares a 58km border: the Somaliland port of Berbera would compete with the entrepôt status of Djibouti, as the main port for the Ethiopian hinterland. It is worth bearing in mind that Somaliland's economy is only marginally smaller than Djibouti's — some $1.4-billion compared with Djibouti's $1.6-billion — and with a coastline and port facility of equal significance. Somaliland has a significant reserve of offshore gas, as well as onshore minerals, and an agricultural potential. In short, it could be a successful commercial rival to Djibouti.

But Somaliland also has an 800 km border with Ethiopia, even greater than its 500 km border with Somalia. Arguably, opening Somaliland to become the entrepôt for Puntland and much of northern Somalia — quite apart from offering another port for Ethiopia — would be part of the solution to gradually introducing stability and prosperity to significant parts of Somalia.

Somaliland, however, has, as we are now seeing, fallen from its well-organized democratic status to a corrupt and, as noted, dysfunctional system, now heavily penetrated by radical Wahhabist proselytizers, who have moved some elements of the society into working with *al-Shabaab jihadists* from Somalia. These problems began with the fall of the Unity of Democratic Alliance (UDUB) Government in the July 2010 elections, which saw the *Hisbiga Kulmiye* (Solidarity Party) candidate Ahmed Mohamed Silanyo take the Presidency. Silanyo, beset by advanced diabetes and probable dementia, had relied increasingly on Minister of Presidency Hersi Ali Haji Hassan, who was essentially an ally and front for the *salafist jihadi* movement, *al-Shabaab*. He had effectively taken control of the Government. Thus ended in 2010 the progress by the outgoing Somaliland Government with the governments of the US, Britain, and Germany for *de facto* recognition.

Kulmiye took power with the help of some Western investors, hoping to win post-election concessions, but this came to nothing. The next scheduled elections were delayed and would not take place in a meaningful fashion unless international pressure was brought to bear. Major state assets — particularly in the Sahel province which includes the port and airport at Berbera — were being sold off to Pres. Silanyo's family and clan, often through foreign intermediaries. This caused a rebellion from the population in the Sahel area, on at least one occasion cutting the highway between Burco and Berbera, thus potentially cutting the ability of Somalians from Puntland, and Somalilanders, to export through the port. This was a clear warning that the opposition to Silanyo was now becoming serious.

One of the great differences between Somalia and Somaliland — because both now have dysfunctional governments — is that Somaliland is still largely self-sufficient, economically, and Somalia is not. Somalia re-entered the International Monetary Fund framework in 2013, but that was clearly a political gesture, and the State is ineligible for IMF funding because the national debt is far too great.

Somalia had an external debt in 2013 of $5.3-billion, some 93 percent of its alleged GDP. In fact, the reality is that the GDP is very artificially calculated, and the debt — which is *not* artificially calculated and had climbed further into 2015 — probably exceeds the real GDP substantially. Foreign grants and trade taxes amount for most of the Government's budget, and 45 percent of that is spent on salaries and wages for government employees. Indeed, given the fact that some 95 percent of Somalia's actual currency transactions are in US dollars is an indication that the State is not faring well. The alleged 3.7 percent economic growth in 2015 was a tribute to foreign sponsorship, not an improvement in local performance.

Some Somali exports are moving, however, with livestock trade to the Gulf Cooperation Council states, and in this regard Somalia, as a "recognized" state by the Arab League, African Union, and United Nations, benefits in that it has literally stolen the trade from Somaliland, which was in the past a major supplier of livestock and hides to the Arabian Peninsula.

It is significant, however, that the Somali livestock exports to Jeddah, Saudi Arabia, during the recent *hajj* went largely through the Somaliland port of Berbera; that included some 1.4-million sheep.

Djibouti

We have already discussed Djibouti extensively in this volume, and there can be little doubt that it will continue to be the focus of significant investment and pressure from major international powers. We may yet see it as a port of call for the Egyptian Navy.

What is significant, however, is the likelihood of change in the political leadership of the State in the coming year or two.

Within this evolving context, the death on July 21, 2015, in the US of Roble Olhaye, 71, the Djibouti Ambassador to the US and Canada (since 1988) and Permanent Representative to the UN, attracted virtually no attention internationally, or even in the Horn of Africa. His death should be expected to cause significant problems for the power base of Djibouti Pres. Ismaïl Omar Guelleh (IOG) and his governing Popular Rally for Progress (RPP) ahead of the 2016 elections.

Amb. Olhaye was, in many respects, the heir-in-waiting and successor to IOG.

Despite the fact that he had been the principal international face of Djibouti for more than two decades — meaning that he was less visible in Djibouti itself — Amb. Olhaye was key to the Djibouti political machine, and his death almost certainly means that Pres. Guelleh should be expected to rely increasingly on his family to project control over the key instruments of State. In particular, Pres. Guelleh's wife, Kadra Mahamoud Haïd, from the Isaaq clan of Somalis, had taken an increasing rôle in policymaking, and one of their daughters is a key advisor to the Presidency. (Pres. Guelleh is also from the Somali Mamassans clan of the Issa, but was born in Dire Dawa, Ethiopia, and educated in Ethiopia).

Sources within the Presidency indicate that former Police inspector and longtime politician Pres. Guelleh intends to change the Constitution — as he did before the 2011 Presidential elections — to enable him to run for another term; this would be his fourth. This makes the US Government nervous, and, indeed, US-Djibouti relations have been fraying because of the growing financial links which Djibouti has with the People's Republic of China.[36]

Eritrea

The aggressive opportunism of Eritrean Pres. Isayas Afewerke seems close to an implosion, but the prospect exists that the 69-year-old former revolutionary will lash out in a final attempt to reassert some regional relevance and to preserve his Government. Certainly, he had, until the recent mineral and port deals, done little to build an economic base in his state.

As discussed in Chapter III, Pres. Isayas was essentially given a new lease of life by the switch of loyalty during 2015 by Saudi Arabia and the United Arab Emirates from Djibouti to Eritrea.

Despite these recent steps, the crisis for Isayas can be seen in the upsurge of Eritreans fleeing across North Africa and into Western Europe, adding to the swelling tide of illegal immigrants there. The situation is far worse than is seen in international media and intelligence reporting. The outpouring of Eritreans comes at a time when Eritrea is ostensibly at peace, unlike Syria, where conflict has driven the population outflow.

The UN High Commission for Refugees (UNHCR) reported in November 2014 that, during the first 10 months

36 See: "Djibouti Takes Center Stage as Horn, Arabian Peninsula, and the Muddied Waters of the Red Sea Are 'In Play'" in *Defense & Foreign Affairs Special Analysis*, August 4, 2015.

of that year, the number of Eritrean asylum-seekers in Europe from Eritrea had nearly tripled: to nearly 37,000. Some 22 percent of "boat people" arriving in Italy during that period were from Eritrea, the second largest number of asylum-seekers after Syrians. The numbers of Eritreans crossing into Ethiopia swelled to 5,000 in October 2014. More than 216,000 Eritreans were already in Ethiopia and Sudan. But 2015 saw these already serious figures skyrocket. Many, until late 2015, were also fleeing across the Red Sea to Yemen and Saudi Arabia.

Isayas' primary bid for survival has been based on conflict with Ethiopia, to destroy the Government there in order to force a compromise which would restore trade from Ethiopia to the Red Sea via Eritrean ports. To achieve this objective, apart from the direct state-to-state conflict which Isayas instigated against Ethiopia in 1998, Eritrea backed numerous armed opposition groups inside Ethiopia. One was the Tigré People's Democratic Movement (TPDM), which the Isayas Government created, funded, trained, armed, and supported for 12 years from 2003 to 2015. On September 11, 2015, however, the TPDM had tired of Isayas and the failure of his endeavors, and fled, *en masse*, into Sudan and thence into Ethiopia.

Eritrean Army units were rushed to the border area, across from the Sudanese town of Omhajer and close to the Ethiopian border, and engaged the battalion-sized TPDM force (est. at around 700 men). The TPDM forces, led by Molla Asgedom, completely destroyed the Eritrean Army force near Omhajer and later at Seq al-Ketir, before heading to Hamdait (all in the Sudan). The TPDM force also suffered heavy casualties, but crossed into northern Ethiopia to be greeted by Ethiopian Government forces at Humera and Dima towns. However, some TPDM groups were still in Sudan, and in the hands of Sudanese security forces.

What is significant is that the TPDM was one of Isayas' most trusted military units, and part of the key to his security. The Eritrean Army is overwhelmingly dependent on forced conscription, for indefinite periods, of unwilling Eritreans, one of the major causes of the outflow of Eritrean men as refugees. Significantly, Eritrean State media mentioned nothing of the defection of the TPDM.

Yemen

Yemen — which had only come into existence in 1990 with the union of the People's Democratic Republic of Yemen (PDRY: South Yemen) based in Aden, and the Yemen Arab Republic (YAR: North Yemen), based in Sana'a' — by early 2016 seemed unlikely to remain as a unified state. Internal divisions and external interference finally had broken down whatever semblance of cohesiveness had been sustained over the preceding 25 years.

The war underway since early 2015 between Saudi-backed Coalition forces supporting the Sunni-dominated Government of Pres. Maj.-Gen. Abd al-Rab Mansour al-Hadi, and the Shi'a-dominated forces supporting former Pres. 'Ali 'Abdullah Saleh, had highlighted the reality that Yemen would probably fragment, sooner or later, into at least two separate states.

December 15, 2015, saw a ceasefire coming into effect in the conflict in Yemen, with the United Nations announcing that negotiations for a possible "permanent end" to hostilities were to begin in Biel, Switzerland. Violations of the ceasefire were reported immediately after it began. It was clear, however, that both sides[37] in the conflict needed to

37 Saudi Arabia had, by late 2015, drawn into its coalition military forces from: Kuwait, Qatar, Bahrain, the UAE, Jordan, Egypt, Morocco, Senegal, Sudan, and Malaysia. Eritrea was also providing military units to the Saudi-commanded operations inside Yemen. The US, UK, France, Turkey, and Belgium were committed to the provision of "non-military" support for the Saudi-led Coalition. However, the US, in particular, was known to be providing military- and intelligence community-sourced targeting intelligence and other intelligence to Saudi Arabia. Iran,

find respite. An initial exchange of prisoners by the combatants occurred after the ceasefire, but a second prisoner exchange failed, leaving Mahmoud al-Subaihi, Pres. Hadi's Defense Minister, still in Houthi hands.

The uncertainty over the future of Yemen as 2016 dawned highlighted the vital significance of the port of Aden, in South Yemen, and al-Hodeidah in North Yemen, to the security of the Red Sea/Suez SLOC. But it also highlights the reality that the international community has forsaken access to Aden's traditional counterpoint across the Bab el-Mandeb: the port of Berbera, in Somaliland. Thus the absence of international safeguards to the actual territories which mark each side of the most strategic waterway in the world should be sufficient to galvanize attention.

Consider the international concern over, for example, Iran's control over the northern shore of the Strait of Hormuz. And at the Bab el-Mandeb Strait, which is arguably significantly more critical to global trade, we now have a vacuum on both the Yemeni and Somali shores of the chokepoint.

There is little doubt that, as of early 2016, the Saudi-led political and military war against the Zaidi Shi'a al-Houthi movement, *Ansar Allah*, in Yemen had served to curtail the establishment of a Government under the Zaidi-dominated Supreme Revolutionary Committee. The question is whether, however, the Saudi campaign can re-unite Yemen, or is the country now more divided than during the era of the two modern Yemens before 1990, when the Yemen Arab Republic and the People's Democratic Republic of Yemen represented two divergent cultures and political systems?

The present conflict in Yemen — that is, since the start of the internationalization of the internal problems in 2014 — has generated a level of displaced persons second only in

Russia, and the People's Republic of China (PRC) were on record opposing the Saudi-led war inside Yemen.

recent years to the Syria-Iraq conflict zone. Some 2.3-million Yemenis have been forced from their homes. It is only a matter of time before the scale of the exodus of Yemenis from their homeland reaches crisis levels for neighboring states such as Somaliland, Djibouti, Ethiopia, and Saudi Arabia. Yemenis had, by 2015, already begun moving through and beyond those immediate destinations and are part of the migratory wave into Europe. As noted earlier, the Yemen conflict has meant that Yemen itself ceased to be a safe-haven for Eritreans seeking to avoid the indefinite penalty of military service. Thus the Yemen conflict has compounded the flow of Eritreans moving further afield for sanctuary.

UN agencies also report that some 20.5-million Yemenis, out of a population of around 25- to 26-million, are now also without safe potable water and proper sanitation, quite apart from food concerns. Damage to the infrastructure from Saudi coalition air strikes and lack of fuel for generators — due to the Saudi naval blockade of Yemeni ports — is hampering normal commercial and social activity. The risk of major health challenges seems likely to rise in the near future, and there is also a risk that these will be carried into neighboring countries, along with refugee flows, as has been happening in Europe with the mounting migratory surge.

A final note on Yemen: the importance of the Yemeni archipelago around the island of Socotra — some 600 miles (970 km) East of the Port of Aden — cannot be missed. Little wonder that the United Arab Emirates, Saudi Arabia's ally in the current Yemen conflict, has made a significant number of aid flights to Socotra in the nine months or more since the Yemen conflict was internationalized, ensuring that the 50,000 Socotrans receive basic food and medicines while the war has severed regular supply ships from Aden. This has helped keep Socotra essentially in the hands of Pres. al-Hadi's faction.

Jordan

The Hashemite Kingdom of Jordan is one of the most strategically sophisticated states in the region, and the most broadly capable military state of the smaller RedMed economies we are addressing here. It is also the biggest of these economies, given the destruction over the past year of the Yemeni economy, which had been equal in overall GDP terms to Jordan's until about 2013.

Jordan has been, and remains, a significantly Western-oriented state, but it has, to some degree, recognized that it needs to find ways to guarantee its sovereignty with the pronounced decline in US or European security support. On October 23, 2015, it signed a new agreement with Russia on counter-terrorism cooperation. Russian Foreign Minister Sergey Lavrov noted: "Under an agreement between His Majesty King Abdullah II and Russia's President Vladimir Putin, the militaries of the two countries have agreed to coordinate their actions, including military aircraft missions over the Syrian territory." Jordanian Ambassador to Moscow Aiad al-Majali said that establishing a "special working mechanism" — understood to be a planned operational facility, based in Amman — to share information on Syrian operations increased military cooperation between both countries to an unprecedented level, noting: "It will not be just in a format of information exchange … (W)e see a necessity 'to be on the ground' as Jordan has a border with Syria. … (W)hen it comes to combating terrorism, we have to" increase Amman/Moscow cooperation. Discussions had been ongoing for some time.

The move reflects the growing disenchantment which senior Jordanian officials have had with what they have felt was the abuse of Jordanian hospitality, in the activities of

US and allied intelligence services in training, arming, and managing anti-Assad *jihadist* fighters from Jordanian soil into the Syrian conflict. As well, Jordanian officials have reacted angrily to an October 7, 2015, statement by US Democratic Party presidential candidate Hillary Clinton which questioned Jordan's political stability and what she described as its "uncertain future".

Jordan, of course, is severely taxed on several fronts, most significantly by the massive influx of refugees from the conflicts in Syria and Iraq, as well as from the Palestinian Authority, Somalia, Sudan, and, reportedly, Eritrea. The overwhelming majority — perhaps as much as three-quarters — of the million or so registered refugees in Jordan in late 2015 were from Syria.[38] Thus, much of Jordan's economic and political focus has, of late, been in containing the crises to its north, rather than focused on the Red Sea. In all of this, and despite hosting US intelligence and special forces units which have been managing *jihadist* radical organizations fighting to overthrow Syrian Pres. Bashar al-Assad, Jordan has quietly been working against the US-backed coalition of Turkey, Qatar, and Saudi Arabia which began, and sustains, the war against Assad. It has been this war — and the ongoing disruption in Iraq which began with the anti-Saddam Hussein war — which has given Jordan most of its grave refugee problems, and which has stirred radicalism within elements of Jordanian society.

But in looking forward, much of Jordan's future strategic strength lies in being able to tap into the Red Sea inlet, the Gulf of Aqaba, to provide desalinated water to transform Jordan's economic prospects. It is understood to be contemplating the use of nuclear power to undertake this.

Russia's Rosatom, on March 24, 2015, signed a $10-billion deal with Jordan to build a 2,000 megaWatt nuclear re-

38 2015 UNHCR country operations profile - Jordan. http://www.unhcr.org/pages/49e486566.html

actor at Amra in the north of the Kingdom, with Russia meeting 49 percent of the project costs. Jordan has been thus far importing 98 percent of its energy, and electricity demand is growing at some seven percent annually. Its energy import costs have amounted to some $3-billion a year. The new nuclear facility would supply some 10 percent of the national needs. [Significantly, in April 2015, Russian Pres. Vladimir Putin, while visiting Egypt, signed an agreement to build Egypt's first nuclear power plant.]

Dermott
&Emery

X

Concluding Remarks and Policy Options

Pamela von Gruber

We have covered an enormous swathe of territory and developments to convey a sense of context to what is happening in the RedMed region, and how it interrelates with Africa, the Middle East, the Mediterranean Basin, and the major global trading powers. Clearly, there is much more we could have discussed, and should have discussed.

It seemed clear, as 2016 dawned, that a decade of massive change was certain to occur in the RedMed nexus, and this was likely to be a key determinant in the global strategic balance moving into the mid-21st Century. It would be unwise, therefore, for any trading power to neglect its interests in that region, or to avoid engagement there to help ensure outcomes most beneficial to its interests. And yet at no time in modern history have the world trading powers — those with a vested interested in unfettered oceanic navigation — been less able to dominate the RedMed.

This, then, is a time when strategic fortunes are there to be made or lost; whole nation-states transformed or lost. And when underlying civilizations may be the most significant rallying points for the societies of the region and its greater terrain.

Let us address some of the conclusions and options

which have emerged from the contributions to this study:

1. The geopolitical centrality of the RedMed region and the Suez/Red Sea SLOC is growing, rather than receding, and this centrality revives the RedMed as a focus of major power competition, particularly between the PRC and the US, but also with other players.
2. The US has been unable to comprehensively arrest the PRC's consistent upgrading of economic, diplomatic, and military encroachment — and therefore strategic rivalry — into much of the RedMed and adjacent areas, and the rise of the PRC's influence in the area must be expected to continue. The question will be whether this Chinese rise will reduce the newly-rediscovered autonomy of the local powers: Egypt, Ethiopia, Israel, and (on the boundary) Iran, in particular.
3. Russia is making significant entrée into the Northern *Bloc* region of the RedMed states, and this could continue to grow.
4. The prospect exists for significant growth in inter-region trade, particularly, initially, between Egypt, Israel, Jordan, Ethiopia, Djibouti, and Saudi Arabia. Ethiopia's regional trade could also be expected to rise significantly in the energy sector if the Great Ethiopian Renaissance Dam (GERD) project is completed as planned.
5. The lower Red Sea zone, in particular, is becoming more, rather than less militarized, and at the Bab el-Mandeb is a zone of conflict or unrest on both sides of the Strait, making it at-risk like no other major shipping chokepoint in the world.
6. Refugee flows within the region should be expected to increase, and some of this will be exported, but some will provide additional stressors on RedMed states. Of particular concern are ongoing population outflows from Yemen, Somalia, and Eritrea, and into Jordan from Syria

and Iraq. These will all have flow-on effects for the Maghreb and Levant and into Europe.

7. Water issues are the great challenges and opportunities for the entire RedMed region, and yet cooperation and interlinking strategies for the region have not been properly attempted; neither in terms of water sharing agreements nor in terms of improving water quality and infrastructure. The limited attempts at Egypt-Ethiopia dialog on Blue Nile water use are still tentative (but significant), as are the overall approaches to water sharing over the entire Nile River Basin. But viable, unilateral water policies are also one of the most significant challenges facing all RedMed states.
8. There is no over-arching mechanism for security coordination between all the member states of the RedMed, a situation which further enhances the prospect for continued instability. And there is no real mechanism for cooperation between RedMed states and some African states which are suffering security problems as a result of activities in, or adjacent to, the RedMed states. Out of area powers, including Turkey, Qatar, and Iran, are engaged in security activities through Sudan, Somalia, and Libya, for example, which have consequences in sub-Saharan zones. Other out-of-region states have joined in military actions in the Red Sea region — whether as part of counter-piracy patrols or as part of the Saudi-led coalition in Yemen — without a comprehensive understanding of the consequences of their actions or the dynamics of the region.
9. The security situation in Somalia shows no sign of being arrested in its fundamentals, and will likely worsen in the long-term if an attempt is made to force Somalia and Somaliland back into a union. However, Somaliland's stability and cohesion is likely to collapse still further un-

less significant international focus is brought to the situation.

10. The international response to the Somalia-based maritime piracy effectively suppressed this phenomenon at a symptomatic level, but it remains uncertain as to whether it can revive if sea-lane safeguards are reduced. Much of the "commercial criminality", as well as ideological radicalism, from Somalia has been re-oriented southward into Kenya, and that may have been in part because ocean piracy was suppressed. The question is now open, however, as to whether, because of the similar reduction of parts of Yemen to chaos or the creation of governance vacuums, there will result a surge of piracy from the Yemeni coastline against international shipping.
11. Egypt and Israel have shifted their electricity systems towards a reliance on natural gas. With offshore gas fields in production and the discovery of new fields in deeper water, both nations face challenges maintaining an attractive investment environment to maintain a growth in energy supplies. With low gas prices potentially lasting to the mid-2020s, there is common interest in ensuring the Eastern Mediterranean is peaceful with low sovereign risk. At the same time, attempts by the US to force Israel into a gas supply relationship with Turkey to save Turkey from the impact of its anti-Russian policies seem unlikely to succeed.

US Policies in the Region

Washington's policies and actions in the RedMed region have been well outlined in this study, but it is worth re-capping the situation as it appeared — in overarching terms — as 2016 dawned. This is particularly important, as senior-level sources in numerous Middle Eastern governments have privately expressed bewilderment at US Government strategies and policies toward the region in the decade or more leading to 2016.

That assumes that an overarching US strategy toward the region still existed by that time. A closer examination of US policies, by early 2016 almost entirely dictated by the Obama White House, gave rise to the belief that no cohesive national goals or policies existed, but rather an *ad hoc* set of actions and reactions which had been largely dictated either by ideological positions, ignorance, whim, or perceived expedience.

This was unique in US history.

In short, the consistent pattern of policies developed over the past century had, by early 2016, been broken up, apart from some of the physical consistencies of legacy military deployments and basing, and by some trade and weapons program commitments. Even there, military deployments had contracted substantially in the past few years, and new US defense systems sales to the region were being lost to suppliers from France, Russia, the People's Republic of China (PRC), Germany, Pakistan, and the like.

In the 18 months until January 2016, the US missed possibly $12- to $15-billion in sales of defense and energy systems in the MidEast, and a range of major new defense acquisitions from non-US suppliers are under consideration by Middle Eastern states. At the same time, some of the US' major traditional allies in the region — Israel, Egypt, and Saudi Arabia, in particular — felt compelled, for their own survival, to turn their back on Washington because of a perception of a divergence in values and goals.

Most US policy officials — especially in Defense — insisted that US commitments and strategies in the region had not changed, but the actions and policies dictated directly by the Barack Obama White House, and mirrored at Secretary of State level, proved antithetical to most states in the greater Middle East, with the exception of Turkey and Qatar. Some regional states, such as Oman, are concerned; others, such as Ethiopia and Djibouti, are now left feeling strategically abandoned.

The sudden withdrawal of US forces from their deployment at the Ethiopian air base at Arba Minch — from where *Reaper* UAV sorties were conducted against *al-Shabaab* in Somalia — was done in September 2015 without forewarning to the Ethiopian Government in Addis Ababa, and kept secret until an Ethiopian website disclosed it in early January 2016. The US had signed a series of multi-year supply agreements with Ethiopian companies to support the base in the weeks leading up to the withdrawal, a firm indication that the decision to vacate Arba Minch was sudden and hastily planned.

The Arba Minch withdrawal coincided with growing US hostility toward the Government of Djibouti — which is strategically integral to Ethiopia's fortunes — and the very pointed siding of US Secretary of State John Kerry with Saudi Arabia and the United Arab Emirates against Djibouti. This resulted in Saudi and UAE strong military commitments to Eritrea (to compensate for the loss of their Djibouti basing in the war in Yemen), another blow to Ethiopian security. But it also coincided with the visit by US Pres. Barack Obama to Addis Ababa to talk at the African Union, where he was accorded a very mixed reception based on his insistence on African states accepting his — Obama's — stance on gay marriage, among other things.

Significantly, although Pres. Obaba's team was warned against such provocations in advance of his Addis and Nairobi visits, most Obama Administration officials did not understand what they had done to offend some of the nations in the region. Even Kerry's support for Saudi Arabia and the UAE in the rift with Djibouti did not win their support for Washington, as both states feared that the US now supported Iran rather than the lower Persian Gulf states. The Iranian Government, however, has been under no such illusions, even among those who supported the G5+1 treaty with Iran to end some of Iranian nuclear weapons programs in exchange for lifting economic sanctions. They, too, see US support for the Saudi coalition against them in Yemen.

The net result has been a bonanza for the PRC, and the deal by Djibouti to welcome a PRC naval base in the country was confirmed and cemented when Djibouti Pres. Ismail Omar Guelleh met in South Africa with PRC Pres. Xi Jinping in early December 2015. This was a strategically successful gathering of African leaders with the PRC leader within weeks of the Indian summit in New Delhi with African leaders.

The US has done nothing of consequence to rebuild its position, which means that the strategic framework in the Middle East and Africa would, within a decade, be profoundly different from the beginning of the 21st Century.

Recommendations and Policy Options

1. There is a need for external players to view the RedMed zone as a comprehensive unit and to end the approach of viewing the Mediterranean and Indian Ocean basins separately. Moreover, the RedMed zone must also be viewed in context with its intimately-linked infrastructures in the Mediterranean and up to the European Union via Cyprus and Greece, and also through the LAPSSET (Lamu Port-Southern Sudan-Ethiopia Transport Corridor project) infrastructure through the Sudans to Kenya and elsewhere in East Africa.
2. Following the recommendation of Gen. Aliyu Gusau Mohammed, it is proposed that the International Strategic Studies Association and the Gusau Institute work to-

gether to create an independent clearing house for informal, confidential exchanges between the member states of the RedMed zone, and to undertake the facilitation of select communications between applicable RedMed agencies and their counterparts in sub-Saharan Africa and elsewhere. Prince Ermias' Water Initiative for Africa has also offered its services to participate in this. We would welcome the initiative and ongoing support of all regional and interested states in transforming this into a reality, so that links can be built with the key regional institutions.

3. Apart from informal, or very discreet Track II approaches to building informational links between the regional states and out-of-region partners, there is a need to focus on bilateral, secure intelligence links between countries experiencing common threats or concerns. Of particular importance would be the exchange of intelligence, for example, between, say, Egyptian intelligence services with those of Nigeria on Muslim Brotherhood linkages and other radical group activities which originate in either the northern states and Turkey or elsewhere in the Middle East, with sub-Saharan groups.
4. There is a need for greater linkage between Indian Ocean and South-East Asian states into the RedMed states so that the patterns of strategic activity can be discerned and acted upon in a timely fashion.
5. We would conclude and recommend that it would be premature for the combined naval task forces and allied naval activities geared to the suppression of piracy and terrorism to be relaxed.

The Sponsors

The International Strategic Studies Association

The International Strategic Studies Association (ISSA) is a global organization, based in the Washington, DC-area, to serve senior national policy officials — elected, uniformed, and civilian — in all countries. It began in early 1972 with the creation of the *Defense & Foreign Affairs* group of intelligence reports for governments, and these were incorporated into ISSA in 1982. It is funded by memberships, subscriptions, and donations, and produces print and on-line *Defense & Foreign Affairs* publications (which are unclassified), and the encrypted-access, governments-only Global Information System (GIS), which monitors 287 countries and territories worldwide, along with key global and regional strategic trends. It has also run the *Strategy* series of seminars for senior officials — up to head-of-state level — since 1972.

ISSA has a worldwide network of field collectors and analysts, and maintains an additional core of experts in its College of Fellows. It also benefits from a distinguished Board of Advisors, which has, for example, included such individuals as HE Field Marshal Mohammed Abdel Halim Abu Ghazala, of Egypt; the Hon. Gen. Alexander M. Haig, Jr., former Secretary of State of the United States of America, and Supreme Allied Commander, Europe; and the Rt. Hon. Sir Geoffrey Pattie, former Minister of Defence of the United Kingdom.

ISSA has a number of analytical focus groups which concentrate on the area of this book, including The Balkan & Eastern Mediterranean Policy Council; the Indo-Pacific Policy Council; and its Middle Eastern Policy Council. ISSA's new study center, the Center for the Study of Monarchy and Traditional Governance, is due to be launched in early 2016.

Websites: www.StrategicStudies. org, www.GISresearch.com, and www.DFAonline.net.

Contact address: PO Box 320608, Alexandria, VA 22320, USA. Tel. +1 (703) 548-1070; facsimile +1 (703) 684-7476.
Email: marketing@StrategicStudies.org.

Key officials: Gregory R. Copley, President; Pamela von Gruber, Executive Director; Yossef Bodansky, Director of Research.

The Gusau Institute

The Gusau Institute (GI) was established in 2014, on the premise that Africa needed a world-class, globally oriented center of philosophy and learning, capable of a unique perspective, based on its location and Afro-centric sources. Africa did not need merely another "African institute" focused on African issues. It needed a body which comprehended the impact of global issues on Africa, and African issues upon the globe. It analyzes the courses of possible action which may successfully address the many human security problems which plague both the continent and the rest of the international community.

Developing long-term working relationships and partnerships with like-minded international entities — corporate, governmental, and non-governmental — is one of the GI's core objectives. Fostering strong international relationships are paramount in ensuring sustainability and future growth.

The GI has an international Board of Advisors, as well as a Board of Governors to assist in its direction and governance.

Website: www.GusauInstitute.com.

Contact address: No. 1 and 2 Dendo Road, Off Ahmadu Bello Way, Kaduna, Kaduna State, Federal Republic of Nigeria. Tel. +234 806 955 7940. Email: info@gusauinstitute.com.

Key officials: HE Lt.-Gen. (rtd.) Aliyu Gusau Mohammed (rtd.), Chairman.

The Water Initiative for Africa

The Water Initiative for Africa (WIA) was formed in 2011 as an international non-governmental organization to help formulate strategic approaches to the use of water, particularly in Africa, and to develop new technologies to purify and deliver stable, ongoing water supplies to African communities for human consumption and agriculture.

The Initiative's goals include the delivery of policies and capabilities to ensure that not only are human consumption for pure, potable water needs addressed as the basic building block of community architecture, but that pure water can revive community-level food production. The common approach to using sustainable technologies to deliver pure water for human and agricultural needs is also designed to eliminate the need for deforestation and the consumption of fossil fuels to purify water, thus reversing deforestation and desertification.

WIA helped to create unique, low-energy water purification systems which can be deployed for sustained periods in areas lacking all other infrastructure. This technology has now begun being deployed in Ethiopia, and other projects in Africa are being developed.

WIA sees part of its function in developing strategic policies on water and waterway useage, whether related to river basin and continental water resources, or related to the uses of oceans and seawater. WIA works on harmonizing sound water-related strategies with social patterns to pre-emptively avoid conflicts and human unrest which are often based on the lack of sound, underlying water infrastructure in the face of changing societal needs.

It is now working with other water-interested groups outside Africa to ensure the provision of new water technologies and thinking for all societies.

WIA is funded and supported through individual donations and directly by The International Society for the Imperial Ethiopian Orders and its members. It is supported by a Board of Advisors which includes, for example: Dr Joan Vernikos, noted scientist and former Director of Life Sciences of the US National Aeronautics & Space Administration (NASA); and former Astronaut, mathematician, and oceanographer Dr Paul Scully-Power, AM.

Websites: www.ArgonautWater.com.

Contact address: PO Box 320483, Alexandria, Virginia 22320, USA.

Key Officials: His Imperial Highness Prince Ermias Sahle-Selassie Haile-Selassie, Patron.

The Authors

Gregory R. Copley: Gregory Copley, an Australian and President of the Washington-based International Strategic Studies Association since its foundation in 1982, is also founder and editor-in-chief of the *Defense & Foreign Affairs* publishing group (since 1972). He is also Chief Analytical Officer for the Global Information System (GIS), which provides intelligence coverage on 287 countries and territories to government users worldwide. He has a long history of writing and analysis on the Indo-Pacific region, and has visited it on many occasions. Three of the 32 books he has authored or co-authored focused specifically on the area: the *Defense & Foreign Affairs Handbook on Egypt*, the book on Ethiopian history and military issues entitled *Ethiopia Reaches Her Hand Unto God*, and the study he co-authored with ISSA Senior Fellow Andrew Pickford entitled *Such a Full Sea: Australia's Options in a Changing Indian Ocean Region.* He has advised a number of governments at head-of-state and ministerial or force command levels for the past few decades. He has also been an industrialist in shipbuilding, aviation, and other fields. He was made a Member of the Order of Australia in 2007, and has been decorated by a number of governments and bodies around the world. He was made a Fellow of the Royal Canadian Geographic Society in 2011.

Lt.-Gen. Aliyu Gusau Mohammed (rtd): Lt.-Gen. Aliyu Gusau Mohammed is Chairman of The Gusau Institute, of Kaduna, Nigeria, and was Minister of Defence of Nigeria until May 2015. General Mohammed formed the Gusau Institute — named after the city of his birth in 1943, and now the capital of Zamfara State — in Kaduna, the city known as the "capital of the north" of Nigeria, to create an African center for strategic studies. General Mohammed is no stranger to the challenges of peacekeeping and security operations in Africa. He has had a distinguished background as an officer in the Nigerian Army, moving directly from his initial military education to be commissioned directly into conflict during the Nigerian Civil War. He was later to become Commander of 9 Infantry brigade, Abeokuta. He served as Director of Military Intelligence, Director of the Defence Intelligence Agency, Acting Director-General of the National Security Organisation, Coordinator of National Security, General Officer Commanding Second Mechanised Division, Chief of Administration at Defence Headquarters, Commandant of the Nigerian Defence Academy, Chief of Staff of the Nigerian Army, three times as National Security Adviser at The Presidency, and concluded his Government service as the Honorable Minister for Defence until his retirement in May 2015. He then assumed his present rôle as founder and Chairman of the Gusau Institute. He has been highly decorated by the Nigerian Government (including Grand Cross of the Order of the Niger) and by governments in Africa, Europe, and Asia.

His Imperial Highness Prince Ermias Sahle-Selassie Haile-Selassie: Prince Ermias is chairman of the Crown Council of Ethiopia and also a Senior Fellow at the International Strategic Studies Association. He is Patron of his charitable organization, the Water Initiative for Africa. That organization's goal is to help Africa devise far better utilization of its water assets, particularly at village levels where water quality is often challenged. Water represents the foundational architecture of all societies. The Water Initiative for Africa is working to introduce new water purification technologies into Africa, without ignoring the strategic realities, and part of its mission is to assist in helping develop a strategic view of water issues.

Ambassador Princeton N. Lyman: Ambassador Princeton N. Lyman, Senior Advisor to the President, US Institute of Peace, began his government career with the US Agency for International Development and served as USAID director in Addis Ababa, Ethiopia, from 1976 to 1978. Subsequently, he was deputy assistant secretary of state for African affairs (1981-1986), US ambassador to Nigeria (1986-1989), director of refugee programs (1989-1992), US ambassador to South Africa (1992-1995), and assistant secretary of state for international organization affairs (1996-1998). From 2008-2010, he was a member of the African Advisory Committee to the United States Trade Representative.

Ambassador Lyman was called back to government by Pres. Obama to serve as the presidential special envoy for Sudan and South Sudan from March 2011 to March 2013. As special envoy he led US policy in the final stages of implementation of the Comprehensive Peace Agreement that ended the Sudan-South Sudan civil war. Outside of government, from 1999 to 2003, Ambassador Lyman was executive director of the Global Interdependence Initiative at the Aspen Institute. From 2003 to 2006, he was the Ralph Bunche Fellow for African Affairs at the Council on Foreign Relations. He has also been an adjunct professor at Georgetown University and at Johns Hopkins School for Advanced International Studies. Amb. Lyman has a PhD in political science from Harvard University. He has published books and articles on foreign policy, African affairs, economic development, HIV/AIDS, UN reform, and peacekeeping.

Andrew Pickford: Andrew Pickford, who was the inaugural Managing Director of ISSA Indo-Pacific, divides his time mostly between North America and Australia, particularly focusing on energy security issues. He has authored or co-authored a significant number of studies, mostly on energy security, but also such landmark books with Gregory Copley as *Australia 2050*, and *Such a Full Sea: Australia's Options in a Changing Indian Ocean Region*, as well as *Energy Security 2.0.*

He has particular expertise with electricity utilities, commodity trends, industry-driven applied research and the reform and transformation of businesses and governments during periods of turbulence. As the author of a range of books, monographs, and articles, Andrew has written and commentated on a variety of issues from agricultural and energy markets to Asian economic trends and the impact of elections on business conditions. He has authored the Economics Section for the two editions of the *Defense & Foreign Affairs Handbook on Pakistan* (2008 and 2015).

Pamela von Gruber: Pamela von Gruber, Senior Fellow and Provost of the ISSA College of Fellows, has been Executive Director of the International Strategic Studies Association since 1986, and Publisher of the *Defense & Foreign Affairs* series of publications since that time. She is also Publisher of the Global Information System (GIS) intelligence database produced by ISSA for government users around the world. She has also been conference organizer for the *Strategy* series of ISSA conferences since 1983. Pamela has also authored reports and articles on emergency disaster response and lectured on the topic.